The
Game Theorist's Guide
to Parenting

THE
GAME THEORIST'S GUIDE
TO PARENTING

How

the Science of

Strategic Thinking

Can Help You Deal

with the

Paul Raeburn and Kevin Zollman

Scientific American / Farrar, Straus and Giroux
New York

Scientific American / Farrar, Straus and Giroux
18 West 18th Street, New York 10011

Library of Congress Cataloging-in-Publication Data
Names: Raeburn, Paul, author. | Zollman, Kevin, 1979–
Title: The game theorist's guide to parenting : how the science of strategic
    thinking can help you deal with the toughest negotiators you know—your
    kids / Paul Raeburn and Kevin Zollman.
Description: First edition. | New York : Scientific American/Farrar, Straus and
    Giroux, 2016. | Includes index.
Identifiers: LCCN 2015036381 | ISBN 9780374160012 (hardcover) |
    ISBN 9780374714406 (e-book)
Subjects: LCSH: Parenting. | Game theory—Social aspects. | Negotiation.
Classification: LCC HQ755.8 .R324 2016 | DDC 641/.1—dc23
LC record available at http://lccn.loc.gov/2015036381

Designed by Abby Kagan

Our books may be purchased in bulk for promotional, educational, or business
use. Please contact your local bookseller or the Macmillan Corporate and
Premium Sales Department at 1-800-221-7945, extension 5442, or by e-mail
at MacmillanSpecialMarkets@macmillan.com.

www.fsgbooks.com • books.scientificamerican.com
www.twitter.com/fsgbooks • www.facebook.com/fsgbooks

Scientific American is a registered trademark of Nature America, Inc.

1   3   5   7   9   10   8   6   4   2

To our favorite game theorists—
our parents and our kids

Contents

The
Game Theorist's Guide
to Parenting

Introduction

In 2005, Takashi Hashiyama, the president of the Maspro Denkoh Corporation, a Japanese electronics company, laid out the terms for what might be the oddest and most expensive children's game ever played.

He wanted to auction off the company's $20 million art collection, which included works by Cézanne, Picasso, and van Gogh. But he couldn't decide which auction house would do the best job—Christie's or Sotheby's. After some thought, Hashiyama announced that he had devised a rather unconventional way to settle the matter: The companies would play Rock, Paper, Scissors. And the winner would sell the art.

Two representatives from each of the companies were called to a Maspro conference room and asked to sit down at a long table facing one another. Each team was given a piece of paper and asked to write one thing: the Japanese word for rock, paper, or scissors. Christie's picked scissors. It was the winner, beating Sotheby's paper.

Most of us would assume that was fair. It was a game of chance; Christie's got lucky. But it wasn't simply luck. Christie's had a strategy. Before the showdown, a Christie's executive consulted two experts on the subtleties of the game: his eleven-year-old twins. "Everybody knows you always start with scissors," they told him. "Rock is way too obvious, and scissors beats paper." After the contest, Sotheby's acknowledged that it had not devised a strategy. It lost several million dollars in commissions as a result.

Rock, Paper, Scissors is a classic example in the study of game theory. Originally developed as a part of mathematics, game theory is the study of how people play games, interact, and negotiate. It deals with strategic thinking in situations where one person's choice of action will affect what the other does in response. Solitaire has nothing to do with game theory: it's just you and the deck of cards. Chess is quite different. Opening with pawn to e4 immediately affects the strategy of your opponent, whose move will, in turn, affect your next move. Those are the kinds of situations that fascinate game theorists.

Game theory was devised by economists and mathematicians, but researchers quickly realized that it had applications far beyond economics. It has become the foundation for industrial-strength negotiations, and it is used by presidents and prime ministers, celebrities and CEOs.

The roots of game theory extend deep into the past, long before its establishment as a science in the mid–twentieth century. One famous example of a game-theory strategy concerns the military figure Kong Ming (also known as Zhuge Liang), who was forced to defend himself against overwhelming odds in a battle in Yangping, China in 149 B.C. It's a twist on the familiar story of the Trojan horse. Both stories involve deception—but in different ways. In the tale of the Trojan horse, the soldiers came out of hiding. In the case of Kong Ming's defense, the soldiers went *into* hiding.

Authorities disagree over whether this is a historical fact or part of Chinese military lore, but the story is too good for us to leave out. Kong Ming was engaged with an enemy, Suma-I, whose forces vastly outnumbered his. Furthermore, Suma-I's warriors had blocked all potential avenues of retreat. Considering the odds, Kong Ming faced certain defeat. He was out of options. Retreating was impossible. Staying to fight was suicide.

At this point, Kong Ming had a brilliant idea. As Suma-I crept forward with his forces, preparing to attack, he stopped. He couldn't believe what he was seeing. Kong Ming had opened the gates to the city of Yangping—and there were no soldiers in sight. (Kong Ming had directed all the guards to hide.) Suma-I saw before him an unguarded city. A solitary figure sat in a tower playing a lute. Suma-I could not understand why Kong Ming would leave the city vulnerable like this. He concluded that it must be a trap, and quickly retreated. Kong Ming had won the battle without throwing a punch. It was a triumph.

The point of this story is that Kong Ming and Suma-I each had to anticipate what the other would think. If Suma-I hadn't stopped to think about why Kong Ming opened the gates, he would have overrun the defenseless city. Kong Ming's actions would have been recorded as a military blunder, and we wouldn't be talking about him now. But Suma-I *did* stop to think. And Kong Ming knew that he would. He knew exactly what Suma-I would think, and he was right. This is why Kong Ming emerges from this story as a military genius.

Game theory did not become a true sensation, however, until 1944, when its principles were spelled out in detail by the brilliant economist, physicist, mathematician, and computer scientist John von Neumann and his colleague Oskar Morgenstern. That was the year they published their groundbreaking book *Theory of Games and Economic Behavior*, which launched game theory as a new science—the science of strategic thinking. Some authorities think von Neumann, an acknowledged genius, was mainly responsible for the breakthrough, and that Morgenstern's role was to goad von Neumann into applying himself to this new field. In any case, their names are now linked as the co-developers of the theory.

One of the most important developments since then came in the early 1950s with the work of John Nash, another brilliant Princeton mathematician whose work on game theory inspired the book and film *A Beautiful Mind*. He published several scientific papers that took von Neumann's and Morgenstern's work much further, and he doubtless would have published many more had he not been diagnosed with schizophrenia a few years later. For a time he left the field, but he ultimately received treatment for his illness and was able to resume his career in the 1990s. In 1994, he was awarded the Nobel Memorial Prize in economics for his work on what are now called *Nash equilibria*, work that explains a lot about human behavior, including how people acting in their own best interest don't always arrive at the best solution. That makes him one of at least eleven game theorists who have won Nobels for their work. Unfortunately, tragedy struck Nash a second time in 2015. He and his wife were killed in a car accident on the New Jersey Turnpike while we were working on this book.

In the years since von Neumann and Nash, game theory has been applied to political science, public health, psychology, and even studies in animal behavior. (Spiders and fish, it turns out, can be excellent game theorists, although they can't be said to think at all. Evolution has equipped them with wonderful strategies that they pursue without knowing why.)

Only recently, however, have game theorists turned their attention to one of the most challenging strategic problems of all—raising children. It's now clear we can put game theory to work in our families. Game theory can help us persuade kids to do their homework, brush their teeth, get out of bed in the morning, and crawl back into it at night. And if we explain the rules clearly to our kids, we can help them get along better with us and with each other—and we can do it without administering any punishment.

Kevin is a philosopher and game theorist at Carnegie Mellon University who studies the evolution of language and the mathematics of social behavior. Paul, a writer, has written two books and numerous articles on parenting, and has five children. We quickly realized, when we got together to talk about this, that game theory has a lot to teach parents. An understanding of game theory and its application to kids and families can help parents avoid arguments, reduce sibling conflicts, and encourage a sense of fairness.

Let's face it—we're up against tough odds. As parents quickly discover, children can be very crafty. Even before they can talk, kids devise strategies to use with their parents. They learn that pointing to a bag of cookies on the shelf or reaching for a toy is a good strategic move—because it usually gets them the cookie or the toy. They engage in the classic maneuver of raising both arms to signify that they want to be picked up. And what parent doesn't comply? Now ask yourself: Who has the upper hand here? Parents need help!

And children are clever, developing strategies that might challenge even some Nobel Prize–winning parents. We should be smarter than our children, right? We've been around a lot longer. But so often they seem to outmaneuver us. When children stay up later than they should, they've outwitted us. When they insist on spaghetti every night for dinner—and they almost always get it—they've defeated us. When they collapse in tears while doing their homework, we've all failed. Nobody's strategy has worked.

In *The Game Theorist's Guide to Parenting*, we will look at how parents and children devise their strategies, where those strategies go wrong, and how we, as parents, can use those strategies to help raise happy, healthy, intelligent children. We're talking about kids who can subconsciously use game-theory strategies to their advantage, but do it with kindness and generosity. The idea is not to set up games in which we are winners and our children are losers. On the contrary, game theory offers us much more than that—the opportunity to help us craft the behavior of our children using games in which we all come out ahead. While game theory can get complicated, in most situations you need to know only three things: the players, their preferences, and what they can do.

You are already using improvised strategies with your kids. That makes you amateur game theorists, whether you're aware of it or not. We will explain how these strategies work, and how you can use them more effectively to encourage your kids to behave in ways that will be good for them—and good for you.

One of the most famous examples of applying game theory to kids is what's known as the Rotten Kid Theorem. It comes from the influential work of the late Nobel Prize–winning economist Gary Becker of the University of Chicago. It's fascinating, partly because it seems to defy common sense. And it's controversial—some economists say the theory doesn't quite fit the facts in real families. But here's how it goes:

You might think that kids who don't care at all about the good fortunes of their family—because they are "rotten"—would not make much of a contribution to their parents and siblings. But if the parents show that they care about the welfare of their rotten kid—despite his behavior—he will soon learn that it serves his selfish interest to treat his parents better—because they will then treat him better. According to the theorem, even rotten kids, in the right circumstances, might be maneuvered into becoming little angels. Or if not angels, then at least less rotten!

As we talk about playing "games" with our children, it's important to remember that we are talking about scientifically based strategies. We're not talking about playing games in the derogatory sense that might prompt one of us to say, "Don't play games with me. I know you didn't do your homework—your teacher sent me a note!" Game theory is not about deception. It's about being smart in dealings with other people—understanding how people are likely to behave, and devising a strategy that will produce the outcome we're looking for. That's true whether we're talking about global political negotiations or about what to have for dinner.

We can use game theory every time we want to divide a scarce resource fairly among our kids, whether it's Twizzlers, LEGO bricks, or time on the iPad. Think of the classic cake-cutting problem: How do we divide a cake fairly when kids have different interests?

The technique for dividing a cake also applies to dividing up time with favorite toys, establishing how much a child's allowance should be, or deciding how to spend a vacation. Game theorists have built careers on the problem of dividing things fairly. Parents who learn how to do that can help to forestall the familiar wail "But that's not fair!" And remember, these little negotiators are clever. A boy who calls a division unfair might mean that he thinks it's unfair. It can also mean that he's setting himself up to have the advantage the next time this comes up: "I didn't get as much candy before! It's my turn now!"

As we started working on this book, we saw that the game theory problem known as the Prisoner's Dilemma has direct implications for the family. So does the theory of rewards. And what about the Ultimatum Game—which deals with the issue of credible and non-credible threats? That one does, too. How many times have we told our kids that we will cancel a trip to the beach if they don't eat their breakfast? They know we're faking, because we want to go to the beach, too—and they know we won't jeopardize our own vacation over a bowl of Froot Loops.

As Ralphie finds out when he asks for a BB gun in the movie *A Christmas Story*, the classic parental admonition—"You'll shoot your eye out!"—can't be beaten. "That deadly phrase honored many times by hundreds of mothers was not surmountable by any means known to kid-dom," he laments. But he gets the Red Ryder carbine-action two-hundred-shot range-model air rifle he coveted, because his strategy plays on his father's sympathies—which trump his mother's fears. Well played, Ralphie! (We wonder whether he'd been reading von Neumann's book . . .)

It's not unusual for children to emerge victorious, as Ralphie did. Too often, we, as parents, feel we're outmatched by people half our size in a strategic game of wits. That's where game theory can help.

We have teamed up to give you the tools you can use to match wits with your kids, to make fair decisions, to stop the squabbling. We know you don't want to fight with your children—and you don't have to.

All it takes is a little thought, a little economics, a little psychology—and a little practice. We're talking about evidence-based parenting. Not fads, not guesses, not tricks. In the best of circumstances, you will create a win-win situation. Not only will you reduce conflict and encourage your children to do what they should, you will also be teaching them strategies for solving problems they will face long after leaving the family and starting out on their own. Game theory works. And everybody wins.

1

I Cut, You Pick

When Paul was a child, he and his younger sister made brilliant use of game theory without knowing they were doing it—and without using a lick of mathematics. They stumbled on what seemed like the perfect way to divide a piece of cake or a candy bar: one would cut, and the other would pick. Perfectly fair, right? Maybe not. Paul quickly discovered, as older siblings often do, that there was a way to maintain an advantage, even while seeming to be scrupulously fair: he insisted that his little sister always cut the cake. She could never divide it perfectly—one piece was always a bit bigger than the other. So by the unassailable logic of I Cut, You Pick, he always got the bigger piece—at least until his sister figured out what was going on.

Cake-cutting is of great interest to kids everywhere, especially those celebrating birthdays. It's also one of the classic problems in game theory. The theory that explains cake cutting is more than two thousand years old, and it's found in ancient texts from all over the world. One of the earliest references appears in the poem *Theogony*, by Hesiod, which dates to 750–600 B.C. In Hesiod's telling, Prometheus—the trickster who stole fire from the gods—aimed to settle a dispute with Zeus by cutting up a great ox into two equal portions. It was a version of cake cutting: Prometheus would cut, and Zeus would pick. "Zeus, most glorious and greatest of the eternal gods, take whichever of these portions your heart within you bids," Prometheus said. That should have led to two fair shares of the cake, ending the dispute.

Prometheus should have been smart enough to know that you don't mess around with a god that can hurl lightning bolts. Instead, he tried to deceive Zeus. One of the portions he prepared was all meat. The other was nothing but bones covered with glistening fat. This second one looked like the bigger and better portion, but it wasn't. The ever-wise Zeus saw through the trick, and refused the bones and fat. Out of vengeance, he withheld fire from mortals. (Prometheus later stole fire in a hollow fennel stalk, Hesiod reports.) Zeus, in addition to knowing how to control lightning, apparently had a little subconscious understanding of game theory. He understood the principle involved—that the person doing the cutting should divide the spoils into two equal portions. Prometheus didn't do that, and Zeus knew something was up. He enjoyed a sweet victory over the scheming Prometheus.

A similar story shows up in the book of Genesis, when Abraham and Lot had more livestock than they could manage, and disputes broke out between their herdsmen. Abraham pressed Lot to end the strife by dividing the land between them.

"Separate thyself, I pray thee, from me," Abraham said. "If thou wilt take the left hand, then I will go to the right, or if thou depart to the right hand, then I will go to the left." Lot chose the plain of Jordan, so Abraham took the land of Canaan. And their dispute ended. Abraham offered Lot either of the two portions, and Lot picked.

The same idea arose yet again in the desperate circumstances during World War II inside Auschwitz, as the writer Primo Levi recalled in *Moments of Reprieve.* "Grigo pulled out a ration of bread and handed it to me together with the knife," he wrote. "It was the custom, indeed the unwritten law, that in all payments based on bread one of the contracting partners must cut the bread and the other choose, because in this way the person who cuts is induced to make the portions as equal as possible."

These are simple cake-cutting problems: Two players. One cuts, and one picks. But cake-cutting can quickly become more complicated if additional players are involved, or if the situation is more complex, as it was in Auschwitz. Most cakes, for example, have different parts. One part might be chocolate, another vanilla. One part might have delicious frosting, while another is covered with hard, tasteless candy flowers. And the outside pieces have a lot more frosting than the inside ones. While we're talking about cutting cake, remember that the principles here apply to many goods and privileges that kids might want to divide fairly, such as time on the computer or television picks.

To explain how the principles of cake-cutting work, game theorists are likely to answer our cake-cutting questions with more questions: What *precisely* do we mean when we say we want to cut the cake fairly? What does "fair" mean in this context?

They have a point. A fair division could mean that the cake is divided into two pieces exactly the same size, with the same amount of frosting. We might feel good about that; it certainly seems fair. But cutting the cake like this doesn't take into account all of the circumstances. The chocolate-loving birthday boy won't feel he's been fairly treated if he gets stuck with the vanilla piece of cake, and neither will his vanilla-craving sister if she's left with the chocolate. Each envies the other's piece, and both are unhappy. Swap the pieces between them, giving each the flavor he or she likes, and the envy disappears, with both kids feeling they've been treated fairly. This is another kind of fairness—a solution that is said to be *envy free*.

Game theorists have been able to prove that even with cakes as complicated as the one we've imagined here, I Cut, You Pick is guaranteed to produce an envy-free division of the cake. For this to be true, of course, the child dividing the cake has to have the motor skills to actually cut it exactly the way he wants to, as Paul's sister eventually learned. This doesn't mean each child gets exactly what he or she wants. It means that each believes his or her piece is as good as the other's. So neither envies the other's piece of cake.

Among the most famous game theorists who have studied this problem are Steven J. Brams of New York University and Alan D. Taylor of Union College, in Schenectady, New York, who describe their work in the book *The Win-Win Solution: Guaranteeing Fair Shares to Everybody*.

Brams and Taylor point out that the notion of cutting a

cake can be extended to many other situations. Some years ago, for example, British and Egyptian archaeologists decided it was time to divide certain archaeological remains between them. The objects were all different, so it was impossible to simply give half of the objects to the Egyptians and the other half to the British. How did they solve it? With I Cut, You Pick, of course.

The British divided all of the objects between two rooms in the Cairo museum. Then the Egyptians were allowed to pick one room or the other. The idea, as with cutting a cake, is that the British would have incentive to make the two collections as comparable as possible, because they knew the Egyptians would pick first.

This strategy doesn't just work for antiquities: parents can use it to divide the labor of raising kids. Suppose that you and your partner have a week of shuttling the kids to band practice, play dates, and doctors' appointments. I Cut, You Pick can work to help divide the labor fairly between the two of you. Mom can separate all of the weekly obligations into two piles that she thinks are equal. She will then be satisfied with either pile. The piles will not be identical; one might represent more yard work and less dishwashing than the other, let's say. If Mom has done her best, they will, however, be equal—in her eyes—in terms of the amount of work required.

When Dad chooses, he picks the pile that he thinks represents the least work for him, or the work he's most willing to do. Maybe he prefers kitchen work to yard work, so he chooses the set of chores that's heavy on washing dishes. Mom cuts, Dad picks.

This kind of inter-parental game theory worked well for Kevin's friends Mark and Tia. One was a night owl, the other a morning person. Mark proposed the following split: One of us should get up with our child in the morning. The other should put her to bed. Tia, the night owl, happily chose to manage bedtime, while Mark, the early riser, was pleased to be handling wake-up. It was a win-win—much better for each of them than, say, alternating bedtime and morning parenting.

Kids can use this trick, too. Suppose your children decide they want to divide a shared box of LEGO bricks, toy cars, or stuffed animals. One child divides the objects into two groups, and the other picks one group.

If this naive use of game theory often happens naturally with kids and adults—as it did, for example, with Paul and his sister—what's the big deal? Why do game theorists find cake-cutting so interesting?

The answer is that this is about more than cake, frosting, and the proper division thereof. Understanding how to divide cake means understanding the difference between an *equitable* division (the two pieces are the same size) and one that is envy free (neither child thinks the other got a better deal). As we've said, the principle here can help with all kinds of situations in which kids need to divide something equally. This can be a useful tool for parents. (And useful for solving problems in government policy and geopolitics, which can seem almost as challenging as raising children.)

Suppose you find yourself, as Paul did recently, standing in the middle of a crowded Toys"R"Us, where an eager store employee is demonstrating a new kind of erasable tablet that you know will never work as well for your kids as it does for him. Your son is interested in the tablet, and he's also trying to raise his voice over the din to ask for a pack of Pokemon cards. Your daughter, meanwhile, wants more Hexbugs to run in the track that already fills up half of her bedroom.

What is a fair division of your resources? How do you cut the cake this time, when the "cake" consists of the tablet, Pokemon cards, and Hexbugs?

Do you spend the same amount of money on each child? You could try that, but what if one Hexbug is more expensive than a whole pack of Pokemon cards? What if the tablet is more expensive than the other two put together? Spending the same on each child won't work. Suppose you ignore the cost and buy each child one toy—one Hexbug for your daughter, and one pack of cards or the tablet for your son. If your son thinks the Hexbug is more valuable than his pack of cards, he might demand another pack of cards—or a Hexbug of his own. If the daughter realizes the tablet is the most expensive gift of all, she might throw the Hexbug on the floor in disgust.

Dividing the resources in your wallet is not the same as cutting a piece of cake—because you are not dividing *all* of the cake—that is, all of the money in your wallet. You are not spending everything you have on Pokemon cards and Hexbugs (we hope). After you've divided some of your money between your children, you still have some left over. If you are dividing a cake into two parts, when it's gone, it's gone.

There is an important distinction to make between cake cutting and toy buying, as we've described them. Dividing a piece of cake is what's called a *zero-sum game*. If one person gets more, the other gets less—by exactly the same amount.

Baseball is a zero-sum game. One team wins (giving it a

+1 in game theory terms) and the other loses (−1). Add them together and you get zero. Buying toys is not a zero-sum game. Both kids can win. (The only loser is you—because you supply the money to fuel this game!) And there is no physical limit on what you can spend. You know—and your children know—that there is more in your wallet than the money you give them to buy Pokemon cards, tablets, and Hexbugs. And it doesn't take long for kids to learn that credit and debit cards might be, in their eyes, a limitless source of free money.

First, let's look at the zero-sum games a little more closely. Zero-sum games (John von Neumann, the father of game theory, invented that term) were the first situations that game theorists tried to explain. And one of the first examples they looked at was chess. One player wins—you could score that as +1—and the other loses (−1). Their scores add up to zero. (If they reach a draw, each scores zero—neither winning nor losing—and the total still adds up to zero.) Von Neumann was also interested in poker, which is another zero-sum game, because the total winnings and the total losses are the same. Every dollar von Neumann lost went into the pocket of somebody else at the table—and vice versa: every dollar he won came from somebody else. Subtract the losses from the winnings, and you get zero every time.

One of the first game-theory principles that von Neumann and his colleague Oskar Morgenstern came up with is what's called the *minimax principle*. They proved that for zero-sum games, minimax play always leads to outcomes in which neither player could improve by switching strategies. Not only that, minimax strategies are "safe." You can ensure that no matter how much smarter the other players are, they can't take advantage of you. The idea is to think about the most you can lose, and devise a strategy to reduce that worst-case scenario. You want to *minimize* the *maximum* you can lose. Hence, minimax! If you begin a poker game with $100 on the table and carry nothing else with you, $100 is the most you can lose. And you can lower the maximum you can lose if you bring only $50 the next time. That's the minimax principle at work—you can't lose any more than what you put on the table. But if you have another hundred dollars in your pocket and you reach for it, you've abandoned your strategy, and the most you can lose now begins to rise.

Cake cutting is a good example of the minimax principle. Suppose Paul's sister is cutting the cake. If she cuts the cake into unequal pieces, she stands to lose more than half of the cake. If she cuts it into two equal pieces, the most she can lose is half of the cake. She has minimized her maximum loss. That's minimax reasoning at its best.

The same reasoning helps us understand what's going on when people make decisions about their health insurance, for example. Some healthy young adults might choose to buy minimal insurance plans, figuring that they are unlikely to get sick, and so unlikely to face high medical costs. That could leave them responsible for thousands of dollars in deductibles and other costs if they do get sick. So why should they buy a better, lower-deductible plan? Because it helps to minimize their maximum loss. If they get sick, they pay lower deductibles and get more comprehensive coverage. Their maximum losses are minimized.

This works only if all of the players are more or less rational. As Ken Binmore, a game theorist at University College London, points out, "game theory can't predict the behavior of lovesick teenagers like Romeo and Juliet, or madmen like Hitler or Stalin." For us as parents, that's not a problem—if we behave rationally, and our kids do, too.

Now, let's be honest about this. Our kids are not always rational. They blurt out things they don't mean. They hurt their own cause by continuing to complain after we give in. And let's be even more honest: We're not always rational either. We blurt out things we don't mean. We're influenced by our emotions as well as our common sense.

Most of the time, however, parents and children behave rationally. We try to act in the best interests of ourselves and the family and try to be even-handed in solving family problems. And given the right encouragement and our intelligent use of game theory, our children will behave in their own best interests most of the time. They might continue to throw temper tantrums now and then, but even temper tantrums, as irrational as they can seem at the time, can sometimes be part of a good strategy for kids. And kids usually know it.

Indeed, children might be more strictly rational than adults when it comes to dividing a cake. In his book *Rock, Paper, Scissors*, the scientist and author Len Fisher describes an experiment he tried at a party where a plate of cake slices was served. When only two were left, he offered them to another guest, who took the smaller of the two pieces. Was this a violation of game theory, which suggests players will maximize their benefits?

He asked the guest why she had taken the smaller piece. "She said that she would have felt bad if she had taken the larger piece," he writes. "The benefit she would have gotten from taking the larger piece (in terms of satisfying her own hunger or greed) would have been more than offset by the bad feeling she would have had about herself for being seen to be so greedy." As it turns out, the game theory prediction was correct. She had not taken the larger slice, which allowed her to have her cake and feel good, too. This is where adults differ from children. Paul can say confidently that his two young boys would always take the larger slice in similar circumstances and feel perfectly fine about themselves. While that makes them good game theorists, he fears it reveals a failure in their moral upbringing—and in his parenting. (This isn't the first time this has crossed his mind.)

The anecdote about the guest who took the smaller piece poses a problem for professional game theorists: *How much* satisfaction and good feeling did the woman get from choosing the smaller piece? What if it was much smaller than the other one? Would she have then made a greedy lunge for the bigger one? Suppose she didn't like the person offering her the cake. Would she decide there was no percentage in taking the smaller piece? What if she hadn't eaten all day?

Game theorists need a way to measure these kinds of nonmaterial reactions. And they do it with a measure they call a *util*—a number that can be assigned to nonmonetary transactions. We will need this idea, because many of our interactions with our children do not involve money. With adults, game theorists can ask people to rate their nonmonetary satisfaction or other considerations by asking them to score their feelings on a scale from 1 to 10. The woman in Fisher's scenario thought the larger piece of cake was a 5 on a 1-to-10 scale, and the smaller one a 4. In terms of her emotional reaction, she scored taking the smaller piece at 8, and the larger piece at 4. The emotional advantages of taking the smaller piece outweighed the value of getting more cake.

That's easy enough. Now let's get back to the problem of cake cutting when the cake is not all the same—not all chocolate, say, or all vanilla. Here another question arises: Does each of the kids know what the other prefers? Sometimes it might be better to be the divider, and it might even make sense to cut the cake into two different-size parts.

Here's how Brams and Taylor break that one down. "Suppose that Ann likes chocolate three times as much as she likes vanilla, but that Ben is indifferent between the two flavors," they write. The cake they are going to divide is 75 percent vanilla and 25 percent chocolate. Ann will divide, and Ben will choose.

Now suppose Ann has no idea what Ben prefers. She must cut the cake into two equal pieces to be assured of a fair division. If, on the other hand, she knows he doesn't care what flavor he gets, she can cut the smaller chocolate piece—25 percent of the cake—away from the bigger vanilla piece, knowing that he will take the big piece—and she gets all of the chocolate, which is what she wanted to begin with. Neatly played, Ann! And it's not a bad deal for Ben, either—he gets 75 percent of the cake. So it's a win-win.

Sometimes knowing the other's preference can be a boon. Brams and Taylor recall the Mother Goose rhyme about Jack Sprat, who "could eat no fat / and his wife could eat no lean." If Jack divides the meat and knows what his wife can eat, he gives himself all the lean and his wife all the fat—and both are deliriously happy. That's what Brams and Taylor call an *efficient* division—each gets the best possible outcome. If Jack divided the meat and didn't know his wife's preferences, he would give each platter half of the fat and half of the lean—and both would be worse off.

With two kids, the analysis is clear. To avoid the scam that Paul perpetrated on his sister, parents can institute a rule that the children take turns cutting, so that one of them doesn't always benefit from an imperfectly divided piece of cake. But what should the parents of three children do? Or what if the cake is supposed to be divided fairly between Mom, Dad, and one or two kids?

This was one of the early dilemmas of game theory. It turns out to be a difficult problem to solve. The mathematics are so complex that it took forty years or so after the publication of von Neumann's book for game theorists to figure it out. The problem has since been solved in a number of different ways by different people.

Suppose you have three kids. The most natural way to divide a cake would be to allow one child to cut and let the other two pick. Fair enough? Not necessarily. Suppose we consider a different problem: three kids dividing a shelf full of toys. The shelf has some toys that appeal to the two younger kids, Jane and Will, and many others they think are boring. Some of those appeal to Tom, the oldest. So here's the plan: Tom divides the toys into three piles. The two younger kids, Jane and Will, can each then pick a pile. That seems fair. Tom wants to make three equal piles, so he won't get shortchanged after Jane and Will pick theirs. If one of the piles is smaller than the others, Tom will get stuck with it. It's just what happens in the cake-cutting example.

If Tom is a game theorist, however, or if he subconsciously thinks that way, he might discover that he has a way to tilt the selections in his favor. He has a different view of the value of the toys, compared to that of Jane and Will. He's too old for some of those toys. So he divides them like this: one big pile of little kids' toys, one small pile of little kids' toys, and one last pile containing all the other toys—including those that appeal to him. (Tom has no reason, after all, to be fair with Will and Jane. Making two unfair piles doesn't hurt him, and maybe he likes to cause trouble for his siblings.) If Jane then picks first, she will take the big pile of little kids' toys, leaving the smaller pile for Will. No fair! If Will picks first, he will choose the bigger pile. No matter who picks first, the other will be unhappy.

What if Jane or Will divides the toys? They might do so in a way that doesn't seem fair to Tom, by making the pile of toys he wants smaller than either of their piles. Now he feels unfairly treated. The others got more than he did, and he will envy them. The division is not envy free, meaning it lacks one of our indicators of fairness.

We can solve this by using what Brams and Taylor call *trimmings*. It works like this: Tom divides the goods into what he thinks are three equal piles. Jane and Will flip a coin to see who picks. Jane wins. She sees one of the piles as the best, another as second-best, and the last as worst. She trims the one she thinks is best until it matches the second best. Will then picks one of the piles. Jane takes the one she thinks is the better of the remaining two, and Tom gets the pile that's left.

But not everything has been divided. What should we do with the trimmings? We could repeat the procedure on the trimmings, although this quickly becomes impractical. After you've tried several times to trim off slivers of cake to play the game this way, you have nothing left except a platter with a mountain of crumbs, assuring that nobody gets a decent slice of cake. And everybody is unhappy. What works in theory, in other words, doesn't always work with children.

There is another issue that affects whether I Cut, You Pick will lead to the best solution. And that issue is: What does each child know about the preferences of the others? As we just saw, knowing what the others want can lead to a better solution than what might happen if the kids have no idea about the others' preferences. Remember Jack Sprat and his wife. If Jack divides the fat and lean equally, his wife will get only half of the fat, which is the only part she can eat. And he gets only half of the lean. It's an equal division, but it doesn't make either Jack or his wife happy. Each gets only half of what he or she wants. There is a more *efficient* solution if Jack and his wife each understand what the other wants. And that solution is to give all the lean to Jack and all the fat to his wife, although this could be rough on her arteries.

Sometimes problems arise when players of I Cut, You Pick games know the other's preferences and act *against* them. Consider spite: the chocolate-loving birthday boy might cut the cake so that he is likely to get all of the chocolate part. If his sister, who has no preference, is nursing a grudge, she might pick the chocolate part just so that he doesn't get it—an act of pure spite. However the boy divides the cake, he is going to feel unfairly treated. I Cut, You Pick has failed him.

Although Brams and Taylor like I Cut, You Pick, they recognize that it doesn't always provide the best way to divide goods. Their research suggests that the best way to divide goods in complicated situations is through a strategy called the *adjusted winner*. It requires assigning point values to objects to be divided.

Many other procedures have been invented for dealing with situations in which more than two people are trying to divide something. Some game theorists have been able to show that solving these problems *requires* really complicated math. Some computer scientists at Carnegie Mellon (where Kevin works) have recently done work on I Cut, You Pick scenarios involving three people or more. They tried to make it simpler by designing—what else?—a website to do the division for you. Called www.spliddit.org, it allows you to fill in the names of your kids, say, and to list the toys they want to divide. Once you've done that, it calculates a solution. Whether or not your children accept the decision depends upon whether they believe that the site is fair. Or it might work by so completely distracting them from the toys that they forget what they were trying to do. (And no, that's not a robust game-theory solution.)

Brams and Taylor discuss another way of dividing goods or privileges that will be familiar to all families. They call it *alternation*, and it simply means taking turns. We all learned it on the playground: two baseball team captains take turns choosing players from the pool of kids who showed up. This has two advantages: The captains don't need to explain their choices; at each turn, they simply pick the player they think is best. The second advantage is that it's easy to understand, and kids think it's fair.

It also has its disadvantages, Brams and Taylor point out. The first is that it might not be envy free. If one player outshines all the rest, the captain who picks first gets an important advantage. If you don't believe it, think about professional sports drafts. Teams that have done poorly get the top pick in the draft, because everyone understands that will help those teams become better. The draft takes advantage of this inherent unfairness in taking turns. If there are three especially talented players, the captain who picks first gets two of them, and the other captain gets only one.

One way to correct for this is through what Brams and Taylor call *balanced alternation*, again familiar to us from sports. In a playoff series, one team gets an important advantage by playing the opening game at home, where, in front of the home crowd, it is more likely to win. To balance that advantage, the other team gets the next two games at home. This is balanced alternation at work. "There is no good reason why strict alternation should be used when balanced alternation is available," Brams and Taylor write.

You might try this when your kids have different ideas about what to watch on television. Let your son pick the first show, and then give your daughter the next two picks. This doesn't mean you have to watch three shows in a row to create a fair division. Your son could pick a show tonight, and your daughter would then get the picks for the next two nights. Then you might find that they are willing to alternate. This could also work for choosing candy, deciding who gets to read with Dad on any given night, or any other circumstances in which they can take turns.

If you have three children sparring over the television, you've entered a realm of higher mathematics in which balanced alternation probably won't work. Brams and Taylor describe three-person scenarios in which Ann, Ben, and Carol try balanced alternation. If they are choosing among three items, they take turns. Brams and Taylor represent that as ABC—Ann picks, then

Ben, and then Carol. But when the three of them are choosing among seven to twelve items, Brams and Taylor come up with this formula to make the choices fair:

ABCCBACBAABC

Ann, Ben, Carol, Carol again, Ben . . . and so on. You are not going to be able to execute a complex strategy like that with children, especially if they are demanding that you step in just as you're putting dinner on the table. Nor will you be able to explain to the kids why the formula is fairer than taking turns. We suggest you invite Brams and Taylor over for dinner, escape to the bedroom with a good book, and let them handle it. Otherwise, just take turns.

You might have noticed that the reason any of this can work with children is because parents are there to enforce the rules. I Cut, You Pick is a good way to forestall arguments and to bring fairness into children's decisions about sharing. It requires that they stick to the rules and accept the outcome. It teaches children one way to avoid conflict, and what it teaches them about sharing is mainly that they should do it because they have to. While Ann is cutting and Ben is picking, Dad is watching. And if this is repeated often enough, the kids should increasingly have faith in the outcomes. If that's the case, Dad no longer has to stand guard.

Our takeaways:

- When dividing cakes, toys, or other fun things between your kids, make sure the division is envy free. This will forestall many sibling arguments.
- The same lesson applies to dividing joys and obligations between parents.
- I Cut, You Pick will often produce envy-free division between two siblings or parents.
- I Cut, You Pick works better when kids (or parents) know what the other one likes and dislikes.
- For more than two kids, try balanced alternation, or you can use www.spliddit.org to help the kids find a fair solution.

2

Don't Cut Barbie in Half!

Kevin's friend Laura wanted to buy a video game system when she was in high school. She couldn't save enough money on her own, so she and her brother decided to combine their allowances and save together. They agreed that once they bought the Nintendo, they'd split time with it evenly. When Laura and her brother had squirreled away enough money, their parents took them to the store to buy the coveted device. They brought it home and took it out of the box, giddy with excitement about all the fun they would have.

But before they turned it on, they realized they had a new problem. They hadn't addressed one question: Who would use the machine first? You can *divide* time: For every hour big brother plays, little sis gets an hour of her own. But how do you divide the *first* time? You can't—someone has to play first. And they both wanted to be that person.

In the last chapter we talked about how to divide objects in fair ways. Cakes can be cut into equal pieces (or at least pieces that feel "equal" to the cake eaters). Time with a favorite toy can be split between the kids. Mom can read a different story on alternating nights. But some items are indivisible. You can't cut the "first time" in half—if Laura played the game first, her brother could not. And vice versa.

As we all know, it's not just deciding who plays first on video game systems that leads to screaming matches in the household. All sorts of problems between kids (whether they're siblings or friends) fall into this category. The new family dog will have only one name, and somebody has to choose it. Only one friend gets to use the cool L.L.Bean Katahdin 35° sleeping bag during a sleepover. The others curl up on the floor with a couple of ragged sheets. If twins disagree about where their birthday party should be, you're stuck. It can be at Chuck E. Cheese's or the neighborhood playground, but not both.

Kids aren't the only ones who face this problem—parents have to juggle parenting with their work and social lives. A friend is having a party, and you can't get a sitter. Who gets to go, and who has the pleasure of putting an overtired child to bed after a long day? Samantha has soccer practice during Jimmy's first piano recital, and both parents want to make the recital, which, unlike soccer practice, is a once-a-year event. Who goes, and who gets stuck with the jiggly iPhone video?

You might look at some of these issues as goods that can be divided. You can "split" them by agreeing to take turns over a longer time span. Perhaps the kids will switch rooms after a year. Little sis can name the dog, and big brother will get to name the cat that you're planning to adopt from the local shelter. George gets the Bean sleeping bag tonight, and his friend Josh will get it next time. But perhaps you've already noticed the problem. Both kids will want the room for the first year. Josh isn't sure if there will be a next time. And who knows if you'll ever get that cat?

The problem of figuring out how to allocate a disputed resource that can't be split is a very old one. The title of this chapter is an allusion to the story of Solomon, who had to adjudicate a dispute between two women, both claiming to be the mother of a newborn. The baby, obviously, had nothing to say about the situation. Solomon recognized that the child should be given to the true mother and not the usurper. But he needed to figure out who the real mother was. To do that, he feigned stupidity and proposed a ridiculous solution—cutting the baby in half—in order to suss out the truth.

The two women believed Solomon was an idiot, and the real mother exclaimed that he should give the baby to the other woman. She did that because she loved the baby and would rather see it be raised by someone else than watch it die. Solomon could then see who really deserved the baby.

It was good for Solomon that he was able to fool the false mother. If she had seen through Solomon's ruse, there would have been a problem. What if she thought about how a real mother might react? Then she could have reacted in the same way. We'll never know how Solomon would have responded to two women both offering the baby to the other.

Solomon's solution worked, but only because the woman making the false claim wasn't that clever. Game theorists worry about situations when everyone is thinking strategically. How do you figure out who deserves the baby when the potential usurper is an expert liar? Hopefully you'll never have to settle an argument over a baby. But sometimes figuring out who gets to drink from the Batman cup feels almost as serious as Solomon's conundrum.

Let's take a look at some of the other options you have when you're faced with the problem of fair division. There is the old standard: Flip a coin. This seems fair, after all. You can't split the birthday party in half, but you can give each child an equal chance of getting her preferred location. If it's heads, we go to Chuck E. Cheese's; if it's tails, we have the party at the playground. That may seem the best you can do. Perhaps if you were feeling a little more creative, you might have the kids play a game where each has an equal chance of winning—Rock, Paper, Scissors might work. This gives the kids the illusion of control, but it runs the risk that a wily kid, like Paul was, might take advantage of a younger sibling.

Game theorists will point out that flipping a coin suffers from two flaws. You've probably already experienced the first one. In economics there is a distinction between allocations that are *ex ante* fair and those that are *ex post* fair. *Ex ante* is just a fancy Latin phrase for "before the event," and *ex post* means—you guessed it—"after the event." Flipping a coin to decide who gets to play the game first is *ex ante* fair, because *before the coin is flipped* neither child has a better chance than the other. Or to put it another way, no child envies the other . . . yet. But after you flip the coin, you can be sure what's to come. One child gets to play first and the other stomps off mumbling about how unfair the world (or Dad) is. That's why flipping the coin is not fair *ex post*. And all the Latin lingo in the world isn't going to make your child feel better about losing the coin toss.

The second flaw with tossing a coin bothers game theorists

even more. Suppose big sister Julia and little Mikey are arguing over who gets to name the new puppy. Both want to name it, but maybe Julia doesn't really value it that much—she's setting her sights on the cat. Now if we flip a coin, Mikey will be crushed if he loses and Julia will be only moderately happy. Wouldn't it have been better to let Mikey name the dog, since he *really* wanted to?

If Julia is a nice older sister, and she sees how much Mikey cares, she might be willing to back off and let him name the dog. But this doesn't always work. Sometimes older sisters aren't so kind—maybe Julia likes to torment her brother. Or maybe she doesn't realize how much Mikey cares about naming the dog. And sometimes children don't even understand how much they care about something unless they give it some real thought.

What would have happened if Solomon had proposed a coin toss? The real mother would have been horrified—because she might lose her child. The usurper would probably be elated, since now she'd have a fighting chance of keeping the baby. But even the usurper could probably have kept her excitement to herself. Half of the time the person who has no right to the child will end up with it!

So, how do we address these two flaws? Game theorists have a favorite solution: auctions. Here's the idea. If you have one item that can't be divided, you want to assign it to the person who desires it most. How do you figure out who that is? Well, you can't just ask—everyone will say, "It's me. Hand it over, because I want it more than any of these other pretenders!" (We're paraphrasing here.) Instead, we ask them to put their money where their mouth is: They must declare how much they would be willing to pay for the item. When it comes to the video game issue, the parents could arrange an auction of the "first play." Laura and her brother could each take turns bidding on the right to play first. At some point, Laura's brother might realize that he doesn't care about being first as much as Laura and he'll drop out, leaving Laura to play the game—for a price.

Auctions have been used as a means to sell all sorts of things throughout history. In A.D. 193, Didius Julianus became Emperor of Rome after winning an auction held by the previous emperor's bodyguards (the Praetorian Guard). Didius Julianus won with his bid of 25,000 sestertii for each member of the Praetorian Guard, and the Guard appointed him emperor. His reign didn't go well: He was executed sixty-six days later. We're guessing the emperorship wasn't worth the price.

Thankfully, most auctions turn out better for their lucky winners. The Stockholm Auction House has been in business for more than 340 years. Christie's, the world's largest auction house, had sales of over £2.7 billion in the first *half* of 2014. Auctions are big business, and they thrive by getting the best prices for merchandise. But if you're not selling an Andy Warhol masterpiece or a global empire (or, we hope, your children), how will any of this help you?

You don't need a fancy auction house or a snazzy website to run an auction. You just need one person to be the auctioneer and two (or more!) happy bidders who are anxious to take home the prize.

When you sit down to run an auction, the first decision you'll need to make is what kind of auction you'd like to run. This may strike you as an odd question—auctions don't differ that much, right? Game theorists can tell you otherwise. There are many kinds of auctions, some invented long before game theory came into existence. Different types of auctions can alter who wins and how much he or she pays. If you want to use auctions to ease family disputes, you will want to be sure to choose the right kind of auction for the problem you're facing.

The most common type of auction is an English auction (sometimes called an *open-cry, first-price* auction). This is probably what you think about when you think about auctions; it's used to sell everything from junk at a swap meet to a Picasso at Christie's. It has two important features: Everyone knows what the bids are, and if you win, you pay what you bid. It's referred to as "open-cry" because everyone knows what someone has bid, and "first-price" because the person who wins pays the amount he or she bid.

Game-theory textbooks almost always recommend a different type of auction called a *sealed-bid, second-price* auction. Game theorists like this type of auction because it's simple to run and easy for the bidders to figure out what to do. In a sealed-bid, second-price auction everyone (secretly!) writes down a final bid on a piece of paper and turns it in to the auctioneer—this is the sealed-bid part. And you won't be surprised to know that the person who bids the most will win the auction. Here's the surprising part: The auctioneer only charges the winner one dollar more than the bid of the *second highest* bidder (this is the "second-price" part).

Here's an example of the sealed-bid, second-price auction at work. Suppose that Korryn, Kim, and Larry all want to control the TV remote. Mom comes in to auction it off (Mom is an economist, naturally). She asks them each to bid secretly. Korryn writes down $10, Kim writes down $12, and Larry writes down

$4. Kim wins the remote because she had the highest bid, but she only has to pay Mom $11—Korryn's bid plus a dollar.

This type of auction is also sometimes called a Vickrey auction after the Nobel prize–winning economist William Vickrey. Economists believed for a long time that Vickrey invented the sealed-bid, second-price auction when he wrote a paper on the subject in 1961.

However, it turns out that these auctions are much older than Vickrey's paper; stamp collectors had been using them long before Vickrey wrote this up. Stamp collecting is almost as old as stamps, and even in the early days—way back in the 1800s—it was common to auction off rare and valuable stamps. For many years, auction houses ran traditional English auctions, where you had to be present to win. Some auction houses would let you hire a proxy bidder who would bid for you. You would send your proxy a letter stating the absolute top dollar you would be willing to pay, and the proxy bidder would continue to bid until your max was reached. After that the proxy would drop out. Proxy bidders weren't cheap—they cost 10 percent of the final price paid.

In 1893, the Massachusetts auction house Wainwright and Lewis decided to do away with proxies entirely and had everyone submit bids by mail. The person who sent in the highest price would pay one cent higher than the second highest price. They could justify this modification to their bidders by arguing that this was exactly the outcome that would have occurred if they had actually held the auction with the proxy bidders. Each sealed bid represented the most that a bidder was willing to pay—which was exactly where the proxies would have stopped bidding. And the winner was the one whose bid topped all of the others. This saved everyone time and money. And thus the sealed-bid, second-price auction was born.

Vickrey didn't come up with this type of auction until almost seventy years later, but he still deserves credit, because he proved a very important fact about these auctions. The best strategy is to write down the maximum you are really willing to bid. There is no trickery here; you say what your top price is. These auctions create an incentive for honesty, and Vickrey proved it.

So why is all this better than flipping a coin? Recall our example auction, where Kim, Larry, and Korryn were bidding on the remote. Supposing that Mom the economist had trained them well, they each bid the maximum they're willing to pay to control the TV. Larry doesn't care much; he's only willing to pay $4. Korryn's in the middle—she'll pay $10—and Kim wants it most of all, valuing the remote at $12. If Mom holds a sealed-bid, second-price auction, Kim wins the remote and pays $11.

Compare this to a coin flip: That would produce jealousy between the kids—the one who was on the losing end of the flip would envy the winner. The auction, on the other hand, produced an outcome that is envy free. While Korryn's still mad that Kim gets to choose the channel, she doesn't want to switch places with her sister. Why not? Because she doesn't want to have to pay $11 for the privilege of the remote. Nor does Kim want to switch places with Korryn or Larry. After all, they can't choose what they watch.

The second problem with the coin flip is that sometimes

the person who doesn't really care about choosing the channel can end up getting the remote. With auctions, if everyone starts out with the same amount of money, the person who values the object most gets it every time. (Larry never had a chance in our auction.) Since everyone has an incentive to bid honestly, the person who bids the most is the person who wants the object the most.

Now, it's one thing to prove it in theory, but will kids really work through the reasoning? We know that adults tend to bid higher in sealed-bid, second-price auctions than economists think they should. This is a puzzle. And economists are busy debating exactly why people are bidding so high. The same effect may well be present in kids, too, and so you might want to try a different type of auction if you find this one does not work with your family.

We've described two types of auctions: the English auction and the sealed-bid, second-price auction. There are hundreds of different types of auctions, and for a while economists thought that they all functioned basically the same way. A famous mathematical theory called the *revenue equivalence theorem* established that if everybody was reasoning perfectly it shouldn't matter how an auction was structured; every auction would yield the same price on average. Alas, experiments have shown that this doesn't quite hold true when it comes to real, live people—different auctions lead to slightly different results.

Here, we'll introduce you to a few more types of auctions so you can try them out. You can conduct your own game-theory experiments to see which ones work best for your kids. And using different auctions can help keep the game fresh and entertaining. You might even manage to teach your children a few critical thinking skills, too. (But don't tell them.) If you confront your kids with different types of auctions, they'll have to learn to think about how to respond to the different ways the auctions are designed.

If you're in a hurry, you might try a variant of the English auction called the *candle auction*. This auction became popular in England in the middle of the seventeenth century, but has since fallen out of favor. This auction works just like the traditional English auction, with everyone raising a hand or calling out a bid in an attempt to win the auction. But there's a twist. In England the auctioneer would light a small candle and let it burn down until it went out. Once the flame disappeared, whoever had the highest bid won even if other buyers wanted to bid more. Of course, you don't have to use a candle; any way of keeping time will do. But be sure that your kids can't tell exactly when the auction will end. Their ignorance prevents them from coming in and bidding at the last second—a familiar annoyance to many eBay purchasers. This type of auction has the advantage of discouraging the slow process of bidding one penny more than the last bid, since the bidders run the risk of running out of time.

A very devious type of auction, which has made appearances at Internet "bargain" sites like QuiBids and Beezid, is called the penny auction. Again, the auction works like an English auction, except you have to pay money to submit a bid (the amount varies by auction site). And every new bid costs you more money—if you bid four times, you pay four times. Auction sites love this one because they make money from all the bidders, not just the winner. So while it looks like the television or concert tickets sold for a huge discount, the auction site still makes money.

Not all auctions are quite so tricky. Kevin's favorite type of auction is known as a *Dutch auction*. This is the most exciting auction, and requires that your kids think carefully about how much their siblings or friends value the objects up for auction. Dutch auctions were introduced in response to an early economic bubble called tulip mania, where the price of tulip bulbs skyrocketed and then crashed again in 1636 and 1637. These auctions are indeed still used in the Netherlands to auction flowers. Dutch auctions are also used by the U.S. Treasury to sell securities, and have been proposed as a way to price companies when they first go on the market.

A Dutch auction is very simple. The auctioneer starts by announcing a price that is ridiculously high—so high that no one would be willing to pay it. Then, each second, the auctioneer reduces the price dollar by dollar. (In the Netherlands, this is done with a giant clock with prices instead of the time written on the face.) This continues until someone jumps up and says "I'll take it." That person pays the announced price and gets the prize.

Some experimental game theorists suggest you use a similar auction called the *English clock auction*. This auction is basically the opposite of the Dutch auction. It starts with a low price—say $1—and keeps increasing the price. Everyone who is willing to pay the announced price raises her hand. As the price goes up, people will drop out. And, once they drop out they cannot come back. Once a hand goes down, it stays down. Eventually, once the price gets high enough, only one person is left. And here is where the auction ends. The one remaining person pays that price.

. . .

Now that you've got plenty of auctions to choose from, you have only two more decisions to make before you become the house auctioneer: What should your kids pay with, and what should you do with the "money?"

All of the theory behind auctions makes one very important assumption about the people who are participating: They all care equally about money. Of course, we all want more money rather than less, for the most part. But economists are assuming a little bit more about how the participants in the auction compare to each other. In order for auctions to be fair in the sense we began with, everyone has to have roughly the same amount of money. Take, for example, Pablo Picasso's *Les Femmes d'Alger (Version "O")*, which sold at auction for $179 million. No matter how much you love Picasso, you're never going to be able to afford that price tag. Now, it's possible that the anonymous buyer loved Picasso more than anyone else in the world. But it's more likely that the outcome was determined first and foremost by how much money the buyer had at his or her disposal.

This can be tricky with kids. You might give your kids different allowances, or maybe one of them hoards the money he gets from Grandma every Christmas. Maybe one of them is saving up for the LEGO Simpsons Kwik-E-Mart, while the other one doesn't have any plans for her money. All of these possibilities might throw a wrench into running your at-home auction.

You could make the payments relative to the difference in allowances. Jimmy the teenager bids in dollars, while Kelly the nine-year-old bids in fifty-cent increments. Fifty cents from Kelly equals one dollar from Jimmy.

But even if your kids have the same allowance, you're still left with the second problem: What do you do with the money you make in the auction? It might seem a bit miserly to keep the dollars and cents for yourself. After all, you didn't get into this parenting gig for the money (and if you did, you need much more help than we can provide—including a new financial planner). A tempting solution might be to give the payments to the kids who lost the auction. Now they can take solace in getting something in return. A game theorist won't like redistributing the money, however, because it encourages your kids to misrepresent how much they are willing to pay. If they know they will get money back for losing, they might bid higher than they would otherwise.

Perhaps you don't care so much about this problem, and if you don't, no worries. If you do, we have bad news for you. Every way of redistributing the money is bad in one way or another. The computer scientist Ruggiero Cavallo and the economist Efthymios Athanasiou have both proposed solutions. Athanasiou's solution involves setting a *reserve price*—if nobody bids above this price, you keep the object up for auction. If you do end up selling it, you decide exactly how much to give to the kids who lost based on a complex formula. This approach ends up being so complicated as to confuse the kids, the parents, and possibly more than a few game-theory graduate students.

But we can deal with both problems at once! Nobody said payments have to be in dollars and cents. So instead of using money, we recommend you try making your kids pay with chores. That's a cost that all kids hate equally. It wouldn't be too hard to implement a chore-based English clock auction. Start by announcing some easy chore: picking up one toy. All who would be willing to pick up a toy in order to get the L.L.Bean sleeping bag should raise their hands. Every hand goes up—that sleeping bag is valuable!

Now make the chore more onerous: cleaning the whole room, perhaps. See if anyone drops out. Keep going and going with more loathsome chores until only one child remains. Not only does this guarantee that the child who cleans the most is the one who wants the prize the most, but it also gets some chores done at the same time. Win-win. And, now that you're auctioning off everything in the house for chores, you have only one thing left to do: Sit back and count your loot.

Our takeaways:

- Auctions are a good way to give out an indivisible prize, like being the first to play with a new toy.
- There are many different types of auctions. Try them all and figure out which one works best for your family.
- Make sure the kids (or parents) all start with the same amount of money, so the auction is fair.
- If you don't want to use money, use something else. We recommend chores.

3

He Got a LEGO Set? That's Not Fair!

Not long ago, Paul's wife, Elizabeth, took their younger boy to the local LEGO store to buy a small Star Wars set. It seemed like a nice way to pass the time on a weekend afternoon while Paul was watching their older son hammer a nice hit for his Little League team at a diamond a few blocks away. The team won the game, and the kids bounced up and down and hugged one another as if they'd just won a trip to the World Series. All that was missing was the dousing with champagne, which, we can agree, would be just a little inappropriate. The ball game was a nice way to pass the time, too, for Paul and his son—until they got home and the ballplayer discovered that he'd missed a trip to the LEGO store. "That's not fair!" he shouted. Was he right? And was this a question of fairness, or of justice? The two can easily be confused.

Paul's first thought was that this was both fair and just. One boy got a LEGO set, and the other got a great win, to which he had contributed. Both kids had had a good day. The older boy didn't get new LEGOs, and the owner of the LEGO set didn't have the priceless experience of winning a ball game, so his trip with Mom seemed fair recompense. On the other hand, Paul had to admit, one boy came home with Star Wars LEGOs and the other didn't. So maybe it *wasn't* fair, thought Paul, already unsure of his initial reaction. Was it just? Had justice been served here? Game theorists—help!

The first thing game theorists will tell you, before they will adjudicate a case like *LEGO v. Little League*, is that fairness is much more complicated than it seems. There is more than one kind of fairness, and none of them are quite the same as what we usually think of as justice.

Game theorists have a story about fairness and justice that makes the difference quite clear. The story comes from one of their own—Sidney Morgenbesser, a professor at Columbia University. During the student uprising at Columbia in 1968, when students were protesting the university's involvement in the Vietnam War, Morgenbesser was arrested for joining a human chain, one of many events in an ongoing series of campus actions. He and others were beaten by the police for resisting calls to break it up.

Later, he was asked whether his treatment was unfair or unjust. "It was unjust, but not unfair," he replied. "It's unjust to hit me over the head, but it's not unfair because everyone else was hit over the head, too." Most of us wouldn't have the clarity and philosophical detachment to describe a police beating as "fair," especially if it had been inflicted upon our heads. But Morgenbesser did, and he had a point. What we're talking about here is not justice, but fairness.

Game theorists like to talk about three basic concepts of fairness. As we saw in chapter 1, one important kind of fairness is what's called the envy-free division of scarce goods, whether those goods happen to be television time or slices of pizza. If your son doesn't envy the piece of cake your daughter got, and she doesn't envy his, the division was envy free. The concept is important to parents. If a division is envy free, then no child will point to another's piece and say, "It's not fair!" Crisis averted!

But equitable cake cutting might not go far enough. If all of the kids at your son's birthday party find your homemade cake a disgusting failure, with a sunken center and misplaced glops of frosting, then nobody cares who gets which piece, or how much—none of them will envy the others, because nobody wants the cake. Just like with Morgenbesser and the protesters, it's an envy-free division all right—but you now have a roomful of angry children on your hands, because nobody is getting any cake. No matter how you cut the cake, *every* child feels cheated.

So: Envy-freeness is enough to seem fair in some situations, but not all. To fix the problem of the unenviable cake, we need a new idea of fairness—one in which each party is as happy as possible. Here's an example: Suppose the cake is half chocolate and half vanilla. You cut it into two pieces, each of which is half chocolate and half vanilla. You give one piece to your chocolate-loving son and the other piece to your vanilla-loving daughter. Neither envies the other's piece, because the pieces are exactly the same. But neither is entirely happy, because each is stuck with part of the cake he or she doesn't want. Something is missing. It's not the best solution.

They could cut up each piece and trade, but a far better solution would be to cut the cake the right way in the first place. That is, cut it into a chocolate piece and a vanilla piece. Then give your son the chocolate piece and his sister the vanilla piece. Voilà! Each is as happy as can be; the division couldn't possibly be better. This reflects a notion of fairness known as *Pareto optimality*. (It's named after Vilfredo Pareto, the Italian economist who described it.) As Paul would put it, a dry martini with a couple of large olives would make him very happy. But it might take an *extra*-dry martini with two blue-cheese stuffed jumbos to make him as happy as possible. A Pareto martino!

Pareto optimality is a goal we will often want to aim for as parents. But before we go on about this Pareto business and the different kinds of fairness, we should step back and ask ourselves a more basic question: Why is fairness so important to us—and to our kids? Why are they outraged when they think a decision or division isn't fair? It's possible to imagine an alternate universe in which kids are happy when they get enough, and they don't care much about what anybody else gets. But that's not the universe we live in. The reason kids get so upset about fairness and unfairness is that a sense of fairness is woven into our evolutionary history. We have this profound sense of fair and unfair because somehow, in our evolutionary past, it helped our ancestors survive. Those who had it outlasted those who didn't.

One way to see this is through a revealing little setup called the Ultimatum Game. In this game, Joe is given a certain amount of money—say $100—and told to share it with Maya. How much Joe shares is up to him. If Maya accepts his offer, they both keep their share. If she rejects it, Joe must forfeit all of the money. The game forces him, in effect, to give her an ultimatum: Take the money I'm offering you, or you will get nothing. But it's a high-stakes game for him, too. If she refuses his offer, he likewise gets nothing.

In classical economics, the solution to this problem is simple. If Joe offers Maya as little as $1—and keeps the other $99—she should take it. Why? Because if she refuses, she will get nothing. Take the buck, for heaven's sake! For an impecunious game-theory graduate student, that dollar could be the difference between starvation and a Styrofoam cup of microwaveable ramen. If Joe believes in classical economics, he will offer her the dollar.

And he will likely be disappointed. People who play the Ultimatum Game in research settings don't accept such stingy offers. If Joe offers Maya $1, or even $10, out of the $100 he was given, she most likely will reject it, studies show. Why? To punish the jerk for offering her such an unfair split. We know this is true because researchers have run the experiments; it's what people do when faced with this situation. And the wise guy who thought he would keep almost all of the money gets nothing either.

Maya is telling Joe, in effect, if you can't do better than that, then you can just go to hell. I don't need your stingy ten bucks. So I'll fix you: I won't get anything, but *neither will you*! Typically, experimenters have found, the person responding will accept the offer only when it is around 40 percent of the total or more. In other words, Maya will take a split that is less than fifty-fifty if it's close enough to seem reasonably fair. The economists who thought she would take as little as $1 were wrong. Those who thought she would settle for, say, $30 or $35 were often also wrong. When this game is played, people want a division close to 50–50, or they walk away.

This works the other way around, too. Joe could offer Maya $99 and keep $1 for himself. That's still better than nothing, and this time *he* gets the ramen noodles. But he doesn't want to do that. He wants to try to maximize his gain, so if he is smart, he will make an offer close to an equal share, in which case the likelihood that he will get *something* is much higher than if he tried to keep too much. He's forced to be fair—or fairer than he thought he would have to be.

This desire for fairness doesn't go away when the stakes get higher. When Lisa Cameron, an economist at Monash University in Australia, performed the Ultimatum Game experiment for three months' wages in Indonesia, she found similar behavior. Even when the money gets big, people sometimes say no.

And remember the woman in chapter 1 who took the smaller piece of cake when given a chance to pick? The classical assumption in economics didn't match up with what actually happened there, either. In that school of thought, the woman should have taken the bigger piece. More is better.

On the other hand, maybe we should be a little careful about bashing classical economics too much. Many economists understand the pitfalls of traditional economic thinking, and indeed, they have come up with a kind of economics to explain why people will sometimes turn down cash. It's called behavioral economics, a not-so-dismal branch of the dismal science.

Behavioral economists take into account nonmonetary factors in economic transactions. The field is a mix of economics and psychology that recognizes how motivations other than money shape these kinds of decisions that seem to defy classical economics. The woman faced with the cake choice took the smaller piece because it was a nice thing to do and she felt that it reflected well on her to do it. And she expected that it would help her in the long run, by showing her to be reasonable and—dare we say it—fair. As with Maya, she rejected the stingy offer because she took some satisfaction in punishing bad behavior. Again, the human psychological impulses trump bloodless economics.

As we said earlier, this is a consequence of the deep sense of fairness in our species. Our sense of fairness—and our kids' sense of fairness—appears to go back to our prehistoric origins. Some scholars have argued that the distinguished philosophers of the French Enlightenment identified the human sense of fairness in the eighteenth century, but that isn't even close to the beginning. If a sense of fairness is apparent in apes, our closest relatives, and even in monkeys, who are less closely related, then we could conclude that a sense of fairness arose way, way back in the mists of prehistory. Our ancient ancestors, we would have to conclude, were pretty good about equal division of wild fruits and spoils of the hunt. If apes, monkeys, and humans all have it, then it presumably arose in the prehistoric ancestor from which we all descended—our most recent common ancestor.

And when did that creature live? Fossil evidence had suggested that humans and apes last had a common ancestor about 7 million years ago. But a genetic analysis of humans and apes published in the journal *Science* in 2014 suggested the split from a common ancestor occurred 13 million years ago. Whichever figure turns out to be right, that creature lived a long, long time ago. And if monkeys have a sense of fairness, too, it goes back even further. Because monkeys are less closely related to humans than apes, their ancestors must have split from ours even earlier. We don't know for sure how long ago the lineages of monkeys, apes, and humans diverged on their individual evolutionary paths. But whenever it was, it looks like a sense of fairness is burned into our genes—and the genes of our children.

Recent studies of monkeys and apes have given us insights into this sense of fairness. One of the first was done in 2003 by Sarah Brosnan of Georgia State University. Brosnan, then a student of the well-known primatologist Frans de Waal, noticed in experiments with pairs of capuchin monkeys that the monkeys didn't like seeing a partner get a better reward. The monkeys were happy to participate in research for a reward of cucumber slices. When they were handed pebbles and then given the chance to swap them for cucumber slices, they happily swapped the pebbles for the food. Good stuff.

But when the routine changed and their partner was rewarded with grapes—a favorite capuchin treat—they became incensed. (No, the reward was not LEGO sets.) Suddenly cucumbers, which had been fine just moments before, were treated as if they were worthless. The monkeys went on strike; they wouldn't play the game. Monkeys who saw their partners gobbling juicy grapes "would get agitated, hurl the pebbles out of the test chamber, sometimes even throwing those paltry cucumber slices," de Waal writes in his book *The Age of Empathy*.

Think about it this way: You offer your son a bag of pretzels, which he likes, for cleaning up his room. He does the work and tears into the pretzels. Then you offer his sister a hot-fudge sundae for the same task and sit her down in front of him while she gobbles up the treat. You know what's coming: *"That's not fair!"*

De Waal writes that Brosnan's initial observation—that the monkeys couldn't tolerate seeing their partners get a better reward—was one of those chance occurrences in science that can easily go unnoticed, but, if followed up, can sometimes lead to important insights. Brosnan and de Waal didn't pay too much attention until they discovered that economists had noticed the same phenomenon in humans. Economists call it inequity aversion. A major component of humans' sense of fairness, it seems, is a dislike of this particular kind of unfairness: uneven or preferential distribution of rewards.

Now let's think again about the Ultimatum Game, and the woman who took the smaller piece of cake. The Ultimatum Game ordinarily demonstrates our dissatisfaction when we're offered too small a share of the kitty, but the woman who took the smaller piece of cake has the opposite kind of inequity aversion: an aversion to getting *more* than someone else. Capuchin monkeys don't show this flip side. A capuchin is perfectly content to munch grapes while another throws cucumber slices around the room.

Our closer relatives, apes, might be more like us. The evidence is spotty, according to de Waal, but he mentions the case of a female bonobo (also known as a pygmy chimpanzee) who received extra rewards (milk and raisins) in the lab. After a while, she began to notice the envious eyes of others upon her, and she refused further rewards. "Looking at the experimenter, she kept gesturing to the others until they too got some of the goodies," de Waal writes. "Only then did she finish hers."

Not all researchers have come to the same conclusions about inequity aversion in animals. Katherine McAuliffe, a developmental and comparative psychologist at Boston College, has done similar studies in animals but failed to find evidence comparable to Brosnan and de Waal's. "I've done it with cotton-top tamarins, capuchin, fish, dogs—and I've never seen it," McAuliffe says. Her work challenges the Brosnan–de Waal findings, leaving the question open, for now, concerning exactly how far back in our ancestry a sense of fairness arose. Either way, however, it's clear that humans have a profound sense of fairness, even if the precise evolutionary history is still unknown.

While that is being resolved, an important question for us is this: Do children exhibit both kinds of inequity aversion? It's pretty easy to see that one child might well be unhappy with pretzels if another gets a hot-fudge sundae. But what about the opposite? At what age do children begin to feel uncomfortable when they get a hot-fudge sundae and another child gets a pretzel? That's a question not for economists or primatologists, but for child development experts.

Consider the case of an older son and a younger daughter. You take your son out to see a movie that's a bit too much for your daughter, and he comes home with a half-eaten bag of popcorn and a movie theater–size box of Skittles. His sister asks for some Skittles and holds out her hand. He carefully places one in her palm, while reserving the remaining ninety-nine Skittles in the box for himself. She expresses her aversion to that "income distribution" just as you'd expect: She wails, "That's not fair!" Your son is playing what looks like the Ultimatum Game, except for one important difference: He doesn't lose his candy if his sister refuses his offer. (Remember that in the Ultimatum Game, if the responder scorns the offer, the proposer gets nothing either.)

Your son is playing what economists call the Dictator Game, in which he is the sole active participant. And his sister has no recourse. She must take the single Skittle, or follow the capuchin monkey example and throw it at her brother. (You might keep an eye out for that.) His sister can't convert this to an Ultimatum Game; she can't deny him his Skittles by refusing his offer. But *you* can! As the parent, you can say to your son, "Do a better job of sharing those with your sister, or I'm taking them away and neither of you will get any." Now it's an Ultimatum Game.

This will get your son's attention. He might not want to be the victim in a Dictator Game—that is, he wouldn't want to be the one forced to accept an offer of one crummy Skittle from his sister. But he clearly has no qualms about unfair splits when he's the one getting more candy. This attitude extends to cake choices, too. If there are two pieces of cake remaining, he's going to take the bigger one every time.

We long for the day when he begins to recognize that getting *more* than a sibling is not fair, and that he will likely get his comeuppance when his little sister shows up with her *own* big box of Skittles. Then he'll be the one on the short end. When they both learn that getting too much is as unfair as getting too little, you might see far fewer sibling quarrels triggered by unfair divisions of Skittles or cake.

When, we want to know, will that day come? When will big brother develop a more sophisticated sense of fairness than his younger sister? When will he be willing to share more than a single Skittle?

Such calculations of fairness are tied up with a sense of morality. Where does our sense of morality come from? De Waal points to a 2001 study by Jonathan Haidt, then a professor of psychology at the University of Virginia, called "The Emotional Dog and Its Rational Tail." He begins the study with what you might call a word problem in morality. A brother and sister, Mark and Julie, are alone in a cabin. They decide that it would be interesting and fun to have sex, relying on birth-control pills and a condom to prevent conception. They go ahead, then decide never to do it again and to keep it a secret.

Was that right or wrong? Most people in Haidt's study told him it was wrong—*so* wrong. And they were sure of it. Haidt challenged their reasons, one by one, until they ran out of reasons. "They point out the dangers of inbreeding, only to remember that Julie and Mark used two forms of birth control. They argue that Julie and Mark will be hurt, perhaps emotionally, even though the story makes it clear that no harm befell them." Eventually, when Haidt has run them through his gauntlet, many say they just *know* it's wrong—even though they can't explain why. As with fairness, then, our sense of right and wrong is so deeply ingrained that we can't always gain access to our own reasoning. We know what we know, and that's it.

Children eventually share that moral sense. They know what's right and wrong, and they're sure of it. Except that their notions of right and wrong change as they grow. When Katherine McAuliffe was trying to find evidence of inequity aversion in capuchin monkeys, it occurred to her that nobody knew how and when children begin to show a sense of fairness. A newborn twin isn't going to know or care whether his sibling is getting more or less milk than he is; he simply wants to be fed. But kids in school know and care mightily about whether someone is getting more snacks, toys, books, and treats than someone else. At what age did this concept of fairness dawn in a child's mind?

It's an important question for game theorists to consider. All of our work adapting game theory to parenting would be a waste of time if we tried to apply it to kids who were not yet able to play the games and have at least an inchoate understanding of what's going on. This is a bit of a sore point for McAuliffe and some other psychologists, who say that game theorists have not done enough to incorporate studies of human development into their game analyses. "We borrow a lot from them; it's unclear how much they borrow from us," she says. "It would be nice if there were models of how cooperative strategies should change during development . . . The assumption is that a human is a human is a human. But humans vary across time and also across cultures."

McAuliffe and a band of like-minded psychologists are starting to use game theory ideas to inform their research into when children exhibit a sense of fairness, and how this sense develops. Alison Gopnik of the University of California, Berkeley, has shown that by age eighteen months infants can grasp that others have minds of their own, and even younger babies are generally more adept at dealing with the world than psychologists once thought. But they are not born with a sense of fairness.

McAuliffe teamed up with Peter Blake of Boston University to try to discover at what age kids develop a sense of fairness. Some research had shown that five-year-old kids, when asked to share a box of Skittles, would give more to themselves than to others, but that around eight years of age, they would begin to divide the candy equally. But according to Blake and McAuliffe, nobody had tested inequity aversion in circumstances in which kids had to pay a cost or make a sacrifice to ensure fairness.

They asked kids to divide candy with an unfamiliar child standing in front of them while parents and others watched. "Children at all ages rejected disadvantageous offers," they discovered. None of the kids was willing to take a smaller share. And that reluctance increased with age. So the kids had an aversion to inequity, right?

Not entirely. When the situation was reversed—when kids got *more* than another child—they were perfectly content! Go to a five-year-old's birthday party that climaxes with the smashing of a piñata, and you will see this result in action: When the thing is finally battered enough by children (and their impatient parents) for the candy to tumble out, five- and six-year-olds dive for the goodies in a bitter struggle to get as much as possible.

Those who emerge with tears in their eyes clutching only a couple of Jolly Ranchers provide an excellent demonstration of inequity aversion—they don't like getting shortchanged, especially if what they really wanted were the Hershey's Kisses. But the kids who emerge with two huge handfuls of all the good stuff show quite the opposite. They're not opposed to inequity—they love it! Inequity is a wonderful thing! Look what I got!

And they are happy to leave it at that, unless their parents insist that they share some of it with the Jolly Rancher kids. The parents might be averse to inequity, but the lucky kids aren't.

Keep watching the piñata scramble, and you will see that the victors are interested in more than what they got—they want to know what everyone else got, too. It's important not only to get a lot of candy, but *to get more than the other kids*. Or, if that's not possible, to get more than the kid from school they really don't like. This isn't just an armchair observation; researchers have found evidence to support it. In a study published in 2014, the Yale University psychologist Mark Sheskin and his colleagues found that five- and six-year-olds "will spitefully take a cost to ensure that another's welfare falls below their own." Another study of seven- to thirteen-year-olds, cited by Sheskin's group, found that the older kids liked scoring well on a certain laboratory test, but they liked it even more if they did better than someone else: "Victory was sweeter when the other child failed rather than succeeded (schadenfreude), and failure was more bitter when the other child succeeded rather than failed (envy)." Future research, they conclude, should not only look at aversion to inequity, "but also the attraction to it."

Perhaps we shouldn't be surprised. We, too, are often attracted to inequity. "Certainly adults engage in constant comparison of themselves with others . . . and there is some evidence that adults show a preference for relative advantage," Sheskin and his colleagues drily observe. This used to be called "keeping up with the Joneses." If we see it in our kids, we shouldn't be too quick to judge.

But do not despair! There is hope for our species. Let's take a closer look at that paper by Blake and McAuliffe.

The paper begins with this scandalously unscientific title: "I Had So Much It Didn't Seem Fair." Experiments since at least the 1980s have shown that adults will go to great lengths to avoid unequal outcomes—not only resisting when they are offered less than another person, but also refusing when they are offered more. So children's attitudes must change somewhere on the way to adulthood. If we want to encourage children to share and to be fair with their friends and siblings—and us!—we need to know *when* it will be effective to apply our parental encouragement. Nothing we say will do much good with five- or six-year-olds, because they feel *quite* comfortable getting more than somebody else.

Blake and McAuliffe set up an experiment in which children could accept or reject an unequal distribution of candy with an unfamiliar recipient. (They made the other recipient a stranger to exclude the context of a prior relationship, which, studies have shown, can prompt children to share more equally—or sometimes, as we've seen with our own kids, less so!) Parents and others were watching, and the kids knew it. The experiment was designed as an Ultimatum Game: Rejecting an unfair split meant both children would get nothing. When eight-year-olds were offered an unfair distribution of candy, even in their own favor, they rejected the deal. They refused all of the candy rather than accept an inequitable split.

"We believe that this is the first evidence that children paired with an unfamiliar peer will forego a benefit for themselves in order to preserve equity, in this case nothing for both children," Blake and McAuliffe write. It kicked in at around age eight. Younger children accepted offers in which they got more than their partner, but by age eight the children stopped doing that. "Children are willing to sacrifice relatively large rewards in order to maintain equity with an unfamiliar peer," the researchers conclude. Eight-year-olds, in other words, are "demonstrating a generalized sense of fairness."

That's the age when we, as parents, can begin to nudge our children in that direction, because they are capable of understanding unfairness—even when it's in their favor. And they care that we are watching.

As McAuliffe notes, game theorists tend to focus on how adults behave, which makes them less interested in children's psychological development. Nevertheless, game theorists have discovered a number of stumbling blocks for establishing and ensuring fairness. And they've devised some interesting solutions to these problems. In experiments on adults, economists have found that the aversion to inequality can quickly evaporate when people find some reason—any reason—to view themselves as more deserving of whatever reward might be there for the taking. In one experiment, researchers randomly dealt eight cards—four aces and four deuces—to two people. The experimenters told the subjects that the aces were worth £10 apiece, while the deuces were worth nothing. The pairs had to agree how to divide the money before they could turn in their four aces. Even though everyone knew the aces were dealt at random, people tended to demand more for themselves when they had more aces.

This is just one example of what the Carnegie Mellon University economists Linda Babcock and George Loewenstein call *self-serving biases*. They say that this tendency to "conflate what is fair with what benefits oneself" is a common psychological error. And it explains why so often businesses and governments, never mind siblings, fail to reach agreements that would clearly benefit all parties.

Now let's check in again with the work of Steven Brams and Alan Taylor, who helped us with the I Cut, You Pick ideas in chapter 1. They remind us of one of the first tricks for fair division that we used as kids in the playground: choosing up sides to play ball. A group of kids settle on two captains—maybe the kids who happened to bring the bat and ball—who then pick their teams from the rest of the group. The technique here is what is technically known as strict alternation: One captain picks a player, then the other, then the first again, and so on. Kids might call it choosing sides. And while it can be brutal for kids who are usually picked last, it's not a bad way of establishing a fair division of resources—two more or less equally balanced teams.

One problem, of course, is who goes first. This can be a big advantage, especially if there is one kid who outshines all the others. If he's not a captain, he's going to be picked first every time, and the captain who picks second will never quite overcome that advantage. "Even if one flips a coin to decide who has the right to choose first, the first chooser may be strikingly and permanently advantaged," Brams and Taylor write. "This bias, determined solely by luck, seriously undermines the fairness of taking turns." Kids who play baseball have a trick for deciding who gets to pick first. One captain throws a bat to the other, who catches it in the middle and holds it there. The first captain puts his hand above the other's, and they alternate, walking their hands up the bat until one of them gets the last hold on it at the end. That one picks first.

Strict alternation has often been used to divide assets in

divorce, Brams and Taylor report. It works reasonably well when all of the objects to be chosen are of roughly the same value—but not so well when things are of unequal value. Imagine a couple making a list of their possessions. At the top is the house—their most valuable possession. The soon-to-be ex-husband wins a coin toss and goes first. And of course he picks the house. The soon-to-be ex-wife picks their second most valuable possession—the car. She would object immediately that this was unfair, and we'd agree.

With a little thought, however, and in the right circumstances, this kind of alternate choosing can be helpful with children. Let's look at dividing books as an example. Suppose a friend or a relative with older kids sends you a box of books their kids have outgrown. And you want to divide them among your children, who insist that they can't share but must each have their own stash. The books are all of roughly equal value to your kids. So taking turns picking them will work. Keep your fingers crossed that each child winds up with the same number of books. Or, if you're clever, you've already done the math and hidden the book or two that could lead to an uneven split. The kids take turns picking. Each winds up with the same number of books. And the split should feel fair to all.

Strict alternation sometimes can be unfair despite our best efforts to set it up fairly. To remedy that, Brams and Taylor introduced "balanced alternation," which we discussed in chapter 1. In its simplest form, it works like this: Your daughter picks a book first. Then your son gets two picks—and your daughter gets one more. It's daughter-son-son-daughter. Each gets two picks, but your son gets two in a row to make up for your daughter getting the first pick. You might try the same strategy if the kids are in a position to divide Halloween candy, or their collection of Star Wars minifigures.

Using alternation can get a bit complicated as the number of items increases. Suppose you have eight books to divide. The first four picks are daughter-son-son-daughter. You can repeat that for the next four books. Or you might want to switch to son-daughter-daughter-son, to make it even. If you have thirty or forty books to divide, you probably want to go back to strict alternation and avoid the need for a spreadsheet to keep it all straight. (If we're talking about dividing goods in a divorce, however, a detailed spreadsheet might be exactly the way to go.)

Still, neither strict nor balanced alternation does everything we would like a fair split to do. When all the choosing is done, one child might envy the other's cache. This solution is not necessarily envy free. And there is no guarantee that the final split will be equitable. So we haven't solved all of our problems yet.

Brams and Taylor came up with a strategy that they think meets all the requirements for fairness—solutions that are envy free, efficient, and equitable. They call their procedure *adjusted winner*. It relies in part on a technique that has been discussed by others: assigning point values to each of the objects in the pool to be divided.

Executing this technique requires stepping up to a blackboard for a little algebra—which we are not going to get into here. Instead, we'll say this: Two teenage siblings want to divide a pile of goods—an iPad, headphones, a couple of guitars, a pile of sheet music, and a beatbox. (This happens to be a musical family.) Each gets 100 points to "spend" on the items. The son might give the best guitar 20 points, leaving 80 to split among the other items. The daughter might assign only 10 points to the guitar but 20 to the iPad. And so on.

In the first step, each gets the items on which he or she placed the most value. The son gets the guitar, for example. The next step is where the math comes in. When the first step is finished, the son might have won stuff worth 65 of his points, and the daughter's items might amount to only 50 of her points. That's not fair. The algebra shows how some items can be subdivided to equalize the winnings. That could be done by, say, dividing the sheet music into two piles, rather than awarding it all to the son or daughter.

Brams and Taylor show how the adjusted-winner technique can be used to resolve international disputes, divorces, and estate divisions. It's complicated, but it's cool. If you're willing to take the time to work it out—and if the stakes with the kids are high enough—this can be worth a try. It works well for diplomats. Why not for kids?

We should remember that what we're discussing about fairness and kids might not be the same everywhere we go. Researchers who have roamed the world introducing other cultures to the Ultimatum Game, for example, found that people in some societies reacted far differently to the Ultimatum Game than American children do. For example, certain cultures in the Pacific Northwest and the west coast of Canada are what are called potlatch societies. They hold gatherings called potlatches in which gifts are exchanged according to rigid rules. The gifts reflect social status and come with an obligation to repay the kindness. So people regularly refuse gifts. When these groups are asked to play the Ultimatum Game, the proposer makes very favorable offers. Remember what happens when Americans play the game. The proposer must offer around 40 percent or so of the hypothetical $100 for the responder to accept. But in potlatch societies, proposers might offer to give away 70 percent of the total—or $70! That's amazing in itself. But what's even more amazing is that responders often *reject* this deal! It puts too much of an obligation on them to respond with equal generosity. The lesson—an important one to keep in mind—is that what is true in, say, American families and with American kids will not be true everywhere. Our shared sense of fairness can be altered by other cultural practices.

Frans de Waal has come to believe that despite all our talk about the moral virtue of fairness and sharing, the sense of fairness evolved because it kept groups from splintering over constant disputes. It seems reasonable to suppose that having cohesive groups served everyone's survival. In other words, we're really interested in fair resolutions mainly to keep the peace.

A cynical view, perhaps. But parents might take a more practical position. Teaching children right from wrong is an important goal that we all share. And reinforcing the sense of fairness is important, too. Over the short term, anyway, most of us would prefer to keep the peace.

Our takeaways:

- Try using the adjusted winner procedure if you're dividing a large amount of goods that differ in quality (books, toys, or teammates).

- Being fair isn't the only thing you want to strive for when you're dividing goods. Try to achieve divisions that make everyone as happy as they can be, without hurting anyone else. Strive for Pareto optimality.

- To teach your kids fairness, make sure they're playing the Ultimatum Game and not the Dictator Game.

- A sense of fairness takes time to develop. Don't worry if your kids don't "get it" right away.

- Help your kids learn to avoid self-serving biases by getting them to think from other people's perspectives.

4

You Can't Be Serious!

It's a story that's been told countless times. Everyone is all packed up and ready to embark on a family vacation. The parents are sitting in the front of the van; the kids are in back. The trip starts out fine, but after the family has made it just a few hours down the highway, screams erupt from the back seat. Julie is tormenting her little brother Stephen. It's just too easy to get Stephen riled up, and Julie can't resist. This time, she's decided to invent a new version of Slug Bug. Every time she sees a silver car, she pokes Stephen so hard he yelps—about two times every mile. Dad tries to get her to stop, but to no avail. Out of frustration, he issues an ultimatum: "If you don't stop, I'm turning this car around! I'll cancel the vacation!"

Julie's shocked. She's been looking forward to this trip for her *whole life*. So she takes Dad at his word and stops harassing her brother. A few minutes pass by. Then Julie has a thought: Dad put a lot of work into planning this vacation. He spent a bunch of money on it already. And just yesterday he was going on about how much he needed the time away. Would Dad really sacrifice his own money, time, and relaxation just because she was poking her little brother? Probably not, she thinks, and pokes Stephen in the ribs.

These stories aren't just for sitcoms. Kevin's friend Desiree watched too much MTV as a young teenager—or at least her parents thought she did. Desiree's parents kept trying to get her to turn off the television, with little success. She'd always sneak into her room to watch MTV when they weren't around. In a last-ditch effort, her parents threatened to cancel the cable. But Desiree saw through the threat. Desiree's father was a huge baseball fan, and their cable company had just expanded its offerings. For the first time in his life, Desiree's father could now watch his favorite team's games on TV! He would *never* agree to cancel the cable. So Desiree ignored the threat and kept tuning in to the latest teenage reality shows. And just as she had expected, the cord was never cut.

The parents in these two stories committed a classic blunder. They made what game theorists call *non-credible threats.* A non-credible threat is a lot like a bluff—it's a threat that is unlikely to be acted on when the time comes—and lo and behold, the two daughters in our stories called their parents' bluffs. You've probably heard the old adage to always follow through— every parenting book tells you that much. But game theorists think about this differently. Commanding you to stick to your guns doesn't make doing it any easier. Instead, a game theorist would suggest that you *design* your punishments with follow-through in mind. You should issue threats that you would be willing to implement (or at the very least that *look like* ones you'd readily act on). This strategy creates consequences that are believable from the beginning and makes it easier on you if you have to follow through.

We'll get to the *how* in a minute, but first a note on the *why*. Because credible threats are the kind that you would be willing to follow through on, it won't hurt you (as much) to impose them. Credible threats make consistency easier; you want to keep to your word. But the benefits of credible threats extend beyond that. When you make your threats credible you'll need to punish your kids less often. Even if the dad in our story had been willing to end the family vacation over Julie's misbehavior, it looked to Julie like he wouldn't—that's why she ignored it. When you make your threats credible you rarely actually *have* to follow through. If Julie predicts that her dad will follow through, then she will modify her behavior accordingly. Of course, not all kids can reason that far ahead. And younger children don't yet have the self-control to act on their reasoning. With older kids, however, you will be handily repaid for making credible threats. Such threats make it easier to punish when you need to, and if done right, they might mean you won't have to.

The idea of credible and non-credible threats was first laid out in game-theory terms by the German economist and Nobel Prize winner Reinhard Selten. Here's the basic idea: a threat is only credible if, when it comes time to follow through, you want to do it.

Selten illustrated this idea with a simple example, called the Chain Store Game, that is a lot like the age-old story of the father and the family vacation. Imagine a small town with only two fast-food restaurants: Ben's Burgers and Sandy's Sandwiches.

They've divided up the territory of the town—Ben is on the east side and Sandy on the west. Sandy decides it's time to expand, so she starts investigating the possibility of opening up a Sandy's Sandwiches location on the east side of town, right in the middle of Ben's territory. She figures that if she opens up a location on the east side she will split that territory's business with Ben, which, if she doesn't cut her prices too deeply, could still be profitable for them both. But obviously it would hurt Ben's bottom line. Sandy also knows that Ben doesn't have the capital to open up on the west side, so she doesn't have to worry about him retaliating in kind.

Ben gets word of Sandy's plan. "Don't do it!" he screams. "If you come into my territory I'll start a price war that will hurt us both. You'll regret messing with me!" But remember, Sandy knows Ben doesn't have much capital—he can't afford a price war. If she opens up on the east side, Ben will have two options: start a price war and go out of business himself (while hurting Sandy in the process) or accept Sandy in his territory. She also knows that despite his bluster, Ben isn't the sort of person to ruin his own business out of spite. So Sandy expands and Ben retreats to lick his wounds.

Game theorists would represent this situation with the picture on the next page. To make sense of this diagram, begin at the top with "Sandy." Here Sandy decides whether to build the new store or to stay out of Ben's territory. If she stays out (follow the line down and to the right) the outcome is written in italics— both continue to get half the town's market. Ben gets the east side and Sandy the west. If, on the other hand, Sandy decides to enter the market (following the line down and to the left), she now

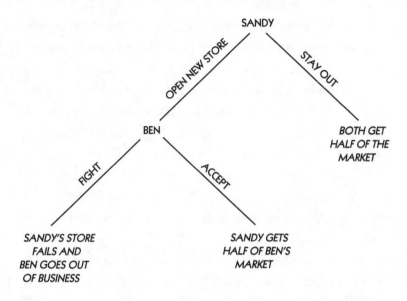

gives Ben a choice. He can either fight Sandy—which results in harm to them both—or he can accept that she's moving into the east side, which results in Sandy gaining some of Ben's market.

Our fathers of game theory, John von Neumann and Oskar Morgenstern, suggested using a reasoning pattern called *backward induction* to solve games like this. Instead of starting at the beginning—with Sandy's move—we should start at the end, with Ben's move. Ignore how you got there for the moment and suppose that Ben is given a simple choice: He can go out of business or he can have a profitable business that splits the east side market with Sandy. Which would he prefer? It seems clear: Ben would prefer to stay in business. So, now we can just assume that's what he'll do in the more complex decision. Now we go to the move right before Ben's move—to Sandy's decision. Since we've figured out what Ben will do, Sandy just has to answer a simple question: Does she want more or less of the market? And her decision is easy.

Backward induction faces some criticism, especially when the number of decisions increases. Interesting paradoxes and problems surround backward induction's application to complicated situations. In our example, there were only two steps to consider. What if there were more—say, one hundred? The reasoning might get too complex. Sandy would have to reason about how Ben would reason about how Sandy would reason. That's already a mouthful, and it's only three steps! But in simple cases with two steps, like the father threatening to turn the car around or the case of *Burgers v. Sandwiches*, backward induction can be very helpful.

Now, when we described the story of Ben and Sandy, we made a number of assumptions that affect how Sandy reasons about Ben's behavior. Notice how the story changes if Ben had the capital necessary to open in Sandy's territory. If Ben could have said to Sandy, "If you open in my territory, I'll open in yours," the story would have been different. Ben would have been threatening to take an action that was good for him—a credible threat. This alters the picture. Ben's threat (moving down to the left) would be better for him than giving in (moving down to the right). A better-capitalized Ben could have used that threat to keep Sandy out of his territory.

Modifying the story in this way represents one method for making a threat credible, and you can use it, too. As a parent you need to threaten with an action that you would be willing to take when the time comes. So rather than threatening to ruin the whole family's vacation, the father in our story could threaten to replace one of Julie's favorite activities with an outing Dad would enjoy. Instead of watching a cartoon in the morning, perhaps the family will head off to an art museum. Now, if Dad is forced to follow through on the threat, Julie will see that he'll actually do it, and maybe it would be better to quit provoking Stephen.

Here's a version of the picture for this scenario:

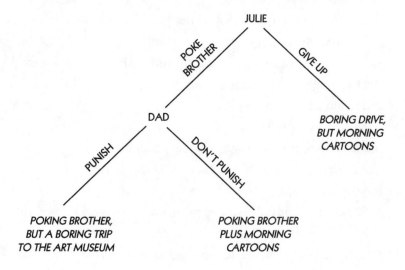

Dad's threat will be credible. If we start at the end the way we did before with Ben, we can see that Dad would be willing to punish Julie. After all, Dad would rather go to the art museum than watch cartoons. So Julie can conclude that if she gives Dad the chance, he'll take them all to a boring museum. The best option for Julie would be to forgo poking her brother and be able to watch cartoons. Because Dad's threat was credible, he didn't even have to follow through.

Sometimes it isn't possible to threaten to do something that you want to do anyway, like going to the museum. Punishing kids is unpleasant, and sometimes it's necessary. But another option is to threaten to take an action that has very low cost to you. Kevin has a friend whose daughters share a room and often talk all through the night. When they won't stop talking, the father threatens to make one of them sleep in a different room. This threat is credible, not because the father has any real interest in making his daughters sleep in different rooms, but because following through will still achieve the father's goal—quiet—at very little cost to himself. And the kids believe he'll do it.

Sometimes the only available threat is something that you *don't* want to do. You may not like taking away TV time or grounding your kids, but sometimes that's the only punishment that will work. How can you still make the threat credible? Game theorists have a solution for this, too: You must find a way to "pre-commit."

This idea found its most clear (and terrifying) realization during the Cold War. President Eisenhower was worried that because the Soviet Union had a significantly larger conventional army, the United States could not deter the Soviets merely by threatening a conventional war. In such a war, the Soviets would trounce the United States. In a policy called New Look, Eisenhower's administration declared that they intended to counter any aggression by the Soviet Union with nuclear weapons.

The military strategist Herman Kahn pointed out a problem with this strategy: The declaration was not a credible threat. Kahn argued that simply threatening to start a nuclear war if the Soviets invaded Afghanistan (for example) wasn't credible. Would Eisenhower really risk global destruction over a small military incursion? The Soviet Union would just ignore such saber rattling. Instead, Kahn floated the idea of a "Doomsday Machine" that would automatically and irrevocably launch a series of intercontinental ballistic missiles—obliterating the world—if the Soviets crossed any of a number of red lines. The Doomsday Machine would take the decision out of our hands. Once armed, it could not be turned off, and even if we regretted our decision, we couldn't stop it. The threat would become credible because we would have irrevocably committed to it ahead of time.

Movie buffs will recognize an almost identical story from the Stanley Kubrick film *Dr. Strangelove*. In the movie, the Soviets build a Doomsday Machine, but decide not to announce it for a while because "the Premier loves surprises." The eponymous character, Dr. Strangelove, points out the absurdity of not telling anyone: "Of course, the whole point of a Doomsday Machine is lost if you keep it a secret!" The similarity here is no accident; Kubrick met with Kahn several times while he was writing *Dr. Strangelove*.

Thankfully, Kahn's suggestion never came to fruition. He recognized a number of the obvious problems with the proposal. It was probably impossible to do, because the computer had no way of learning about global events. We could not tolerate the smallest risk of error. It would be obscenely expensive. And, if anyone got wind that we were building a Doomsday Machine, they would rush to try and build their own, creating a second, and even more terrifying, arms race.

As crazy as Kahn's idea was, however, there was a grain of game-theoretic truth in it. In fact, the principle of precommitment has been used throughout history. In the third century B.C., a Chinese warlord and rebel, Xiang Yu, crossed a river to battle Zhang Han, a general of the ruling Qin dynasty. Yu faced overwhelming odds. He recognized that his troops might expect him to retreat if the battle went badly, and that this might cause them to hang back or fight with less zeal. Yu knew that he had to find a way to make his troops fight as hard as they could if they were to have any chance of victory. So, Yu sank all of his ships and left his troops with only three days of food. No retreat or delay was possible; they had to fight, and the only hope of survival was to fight as hard as they could. Yu's strategy worked. With nothing to fall back on, his troops won the battle. (It has been said—although it's debated—that this strategy was used again by the Islamic general Tariq ibn Ziyad in the eighth century A.D. and again by the Spanish Conquistador Hernán Cortés in the sixteenth century.)

Although it may sometimes feel as though you're waging war on your kids, you probably aren't going to scuttle any ships. But the idea of pre-commitment embodied in these stories can work in mundane situations, too. Consider the story of Ben's Burgers and Sandy's Sandwiches again. Suppose that Sandy is threatening to open in Ben's territory and Ben can't afford a price war. Now suppose that Ben is about to head out of town on vacation, and then suddenly he's struck by an idea. He calls a lawyer and draws up a contract. If Sandy opens the store while Ben is away, the lawyer is obligated to start a price war, regardless of how it would affect the business. Ben pays the lawyer in advance, so the lawyer has every incentive to follow through on the contract. Then—and this is important—Ben sends a copy of the contract to Sandy. (Remember Dr. Strangelove's admonition: You've got to tell somebody about it!)

Ben has just built his own little Doomsday Machine. Here's how that changes our Ben and Sandy diagram:

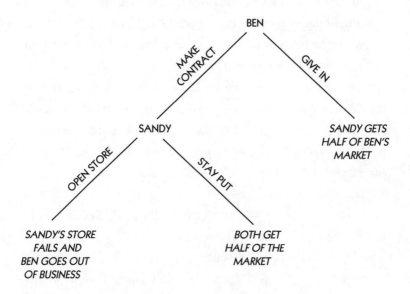

Notice, Ben gets to move first this time—he beat Sandy to the punch. Let's walk through backward induction using this new example. Like before, we start with the last move, but this time that move is Sandy's. She has to decide whether to open a store that will fail or skip it and make do with half of the market. Obviously, she would rather not open a store that has no hope of surviving, so she won't bother. A step back in time, Ben must choose between executing the contract with his lawyer or throwing it in the trash. Obviously, he'd like to keep Sandy out, and so he and his lawyer sign the contract. He can now show Sandy that the threat is real.

So how can you build your own Doomsday Machine? Without knowing it, you probably already have. Many schools have strict academic requirements for playing sports or participating in the band. So when Mike, the high school's star baseball player, fails economics, the coach and Mike's parents have no choice but to suspend him from the next game. The punishment is automatic.

While you may have already benefited from pre-commitment, now you'll be armed with the background on its game-theory roots. And you'll be able to come up with new commitment strategies of your own. For example, making a promise to another child or parent can work as a Doomsday Machine. Suppose the father at the beginning of this chapter had told Julie that if she poked her brother again, then her brother, Stephen, would get to choose the music for the remainder of the trip. Even if the father hates Stephen's taste in music, by making that promise in front of Stephen, Dad the Strategist built a Doomsday Machine. Julie knows that if Dad goes back on his threat, he'll get an earful from Stephen. (Parents or caregivers can use each other as Doomsday Machines, too. Mom can tell Dad that if Stephen misbehaves Dad gets to choose the channel for that night's hour of TV. If Mom makes that promise in front of Stephen, he now knows that Mom has to keep her promise and punish him—even if Mom hates Dad's favorite TV show, too.)

In addition to making threats that you would like to follow through on, and creating ways to pre-commit, there is one more way to ensure that your threats are credible. It's developing a reputation. Interactions between parents and their children are rarely just one-offs: They are repeated. And establishing a reputation for making good on your word can make threats credible. You have to carry through on your threats today so that tomorrow's threats are credible. This idea may be as old as social interaction, but it was first articulated in modern form by the philosopher Thomas Hobbes (only a philosopher such as Kevin would call 1651 "modern"). In his famous book *Leviathan*, Hobbes argued that people who break their promises will come to regret it, because they will be excluded from society and will ultimately lose far more than they could gain by being untrustworthy.

We'll spare you the equations, but trust us: We can even give Hobbes's point about reputation a precise mathematical form, and we can prove it with mathematical certainty—that threats can become credible when two parties have dealings over and over again. And analyzing the costs and benefits of threats and reputation in mathematical terms highlights an important caveat: The gain from future credibility must be large enough to outweigh the short-term costs of following through. So, if Dad is going to issue a credible threat to Julie, two conditions have to be true: (1) the cost of carrying out the threat has to be small; (2) the benefit to his stellar reputation as a man who means what he says must be large.

Now we can see why Dad's threat to turn the car around didn't work. Canceling the vacation would come at a large cost, and he might not be willing to pay it just to develop a reputation as a disciplinarian. It just wouldn't be worth it to follow through—and Julie knows it.

Game theorists have identified other important features of punishments. Punishments must have a cost for the kid that's higher than the benefit gained from behaving badly. This may seem obvious, but it can be easy to screw up, because we normally think about punishments and rewards in terms of how we, the parents, would react. How would an adult respond if she had to pay a price to poke someone painfully in the ribs? Well, she probably wouldn't poke anyone (at least, not that hard) to begin with.

But game theory tells us not to think that way. Instead, we must always consider the costs and benefits *from the perspective of the person who has a choice to make*—and parents can easily go wrong by considering their child's choices from the grown-up's perspective. Kevin's parents made just that mistake when he was having trouble staying focused during high school. Kevin was bored by his sophomore history class, and instead of just paying attention and getting through it, he drifted off and thought about the computer game that had recently caught his attention. As a result of his daydreaming, Kevin ended up failing the semester. Kevin's parents were mortified. To make it up, Kevin had to attend night classes, and Kevin's parents thought that the loss of his free time would be sufficient to deter him from failing any future classes. But, alas, they were wrong. Kevin loved the night classes! They let him work through the material at his own pace, so he breezed through a semester in a few days, took all the tests, and passed the class. He realized he could fail all his classes, take night classes, and end up with more time to himself than if he worked hard in his regular classes.

Kevin's parents were forced to consider supplementing the school's punishments with their own. In the end, it was high school debate—with its academic requirements—and more engaging classes that got Kevin back on the right track. Even his parents admit now that he turned out okay (for the most part).

The consequences of a punishment need to be evaluated from

the perspective of the person being punished. What Kevin's parents missed was that he would rather daydream through his regular classes and catch up at night (in less time) than pass the class itself. If they had foreseen that, they could have included some additional incentives to keep Kevin on track.

One more important feature of punishments—when they become necessary—is the time that transpires between your kid's misbehavior and when he or she must suffer the consequences. The problem with young children is that the future seems so far away for them that they don't often think about what it will be like. The punishment might as well be happening to someone else. Even older children have a difficult time thinking a few years out. Declarations like "You won't get into college!" or "You'll never get a good job!" just won't register. Children think about the future in very different ways from adults. We'll discuss how to adapt to the way your kids see the future more in chapter 8.

Everything we've just said about punishment goes for rewards, too. In order to motivate a child to follow the rules or perform well in school, rewards need to satisfy the same criteria as punishments. The reward must be credible—promising your child a pony won't work, at least not more than once—and it must be sufficiently good *for the child* to outweigh the benefit of behaving badly. (How much does your child relish tormenting a sibling?) And, like punishments, rewards that won't appear until much later are likely to be ignored. While your daughter might trade ten minutes of a video game for an ice-cream cone she gets immediately, she might not trade those same ten minutes for an ice-cream cone that she'll get next week. While we adults might be happy to forgo a benefit now for a reward next week, our children may not.

In fact, traditional game theorists rarely distinguish between punishments and rewards, because in many respects they are just two sides of the same coin. Does Dad punish his daughter for bad behavior by refusing to play toss? Or does he reward her for good behavior by agreeing to play toss? But one important difference between framing an action as a punishment or reward has been pointed out by researchers in the part of game theory called behavioral game theory. The psychologists Amos Tversky and Daniel Kahneman discovered that people often think about "rewards" and "punishments" differently. In a famous experiment, Tversky and Kahneman told the following story to a group of students who volunteered to be in their study:

> Imagine the U.S. is preparing for an outbreak of an unusual Asian disease, which is expected to kill 600 people. Two alternative programs to combat the disease have been proposed. Assume that the exact scientific estimate of the consequences of the programs are as follows:
>
> If Program A is adopted, 200 people will be saved. If Program B is adopted, there is a 1/3 probability that 600 people will be saved and a 2/3 probability that no one will be saved.

Kahneman and Tversky then asked the group which program they preferred. Stop for a second and think about which one you might choose. A large majority of the students preferred Program A (the one that was guaranteed to save two hundred people) over Program B, which might save everyone, but could also totally flop. If that's what you chose, too, congratulations—you're in the majority. But don't fret if you chose Program B. There is nothing intrinsically wrong with your choice—it just appealed to fewer people.

Kahneman and Tversky then recruited another group of students and told them the same story about the virus, but this time they offered them a choice between two apparently different programs:

If Program C is adopted, 400 people will die. If Program D is adopted, there is a 1/3 probability that no one will die and a 2/3 probability that 600 people will die.

Which one would you prefer? When the researchers presented the choice *this* way, a large majority preferred program D. But notice: A and C are *exactly* the same programs, and so are B and D! In the first case they're described in terms of who is "saved," and in the second they're described in terms of who will "die." What Kahneman and Tversky found was that when we think in terms of good outcomes—saving other people—we want to opt for a sure thing and not take a risk. But when we think in terms of bad outcomes—people dying—we are willing to take risks, at least if there's some chance of avoiding the bad outcome completely.

Kahneman and Tversky's results have been repeated over and over, including with children. The game theorist Joshua Weller and colleagues examined how children value risk compared to adults by offering them choices between a sure thing (a quarter) and a risky one that might yield several quarters or no quarters at all. Young children would take risks regardless of whether the possible outcomes were described in terms of gains or losses. As children got older they started looking more like adults, preferring not to risk a good sure thing. By the time they reached eighteen, their choices followed the same pattern as in Kahneman and Tversky's studies; they were more willing to take risks to avoid losses than to secure gains.

To see what all these studies say about punishments and rewards, let's consider another example. Tim is sixteen and he just got his driver's license. He'll use any excuse to take the family car out for a spin. Like any teenager, he also likes to go out with his friends and get into all sorts of mischief—nothing dangerous, but enough that Mom and Dad occasionally get angry phone calls from a neighbor who has recognized Tim chucking rolls of toilet paper onto his house. Of course, sometimes Tim doesn't get caught and gets away with it. If Mom threatens to punish him by taking away car privileges, he's facing a punishment—a "loss" in Kahneman and Tversky's theory. Their study suggests that he might be willing to take the risk of being punished. "Maybe I won't get caught!" he might think. But if he's facing what seems like a gain—if Mom "rewards" him for not causing trouble by letting him drive the family car—then he might not take that risk. "Maybe I *will* get caught!"

Since you can often recast your punishments in terms that make them sound like rewards, we're suggesting that you approach your children with rewards more often than with punishments. Instead of threatening to nix a promised ice-cream cone if he makes a fuss at the restaurant, offer him an ice-cream cone if he's good. By framing the prospect in terms of gains, you may just get him to stop acting as if he won't get caught.

All these strategies won't make your children into little angels overnight, but by using these lessons you'll make the inevitable conflict easier on everyone, and, with a little luck, reduce the times when you have to be the bad guy.

Our takeaways:

- Make your punishments or rewards harm you as little as possible. When possible, make a punishment at least appear to be an action you would be happy to carry out.
- When possible, make irrevocable pre-commitments.
- Make sure your child sees that you're pre-committed (the Dr. Strangelove principle).
- Don't simply rely on your reputation (although this can help). But don't ruin your reputation by not following through.
- Think about punishments and rewards from the perspective of your child, taking into account how your child thinks about the future.
- Where possible, frame punishments in terms of rewards.

5

The Dog Ate My Homework

Like almost every middle-schooler in the world, Kevin hated homework. He would do anything to avoid completing worksheets and practice problems. Sitting in front of the tube always seemed more captivating. Kevin's parents, on the other hand, wanted to be sure he finished his assignments. At first, they handed down a decree: Kevin would be required to do his homework before doing anything else. Every afternoon when Kevin came home from school, his parents would ask him if he had any homework. At first Kevin answered honestly; he would tell them exactly what his assignments were. Then he would head off to his room to complete his work. But after a while he figured out an alternative. He could lie and say that he had just a little homework, and get to watch TV earlier.

We shouldn't blame Kevin's parents for his dishonesty, but they did create conditions that encouraged it. The rule they established—that Kevin had to do his homework—required that Kevin be honest in order for the policy to work. Kevin didn't want the rule to be enforced, and thus the seeds of his dishonesty were sown.

Could Kevin's parents have done anything differently? Could they have set up rules that would have created an incentive for honesty instead of deception? Keeping kids honest is difficult, as you've probably already discovered. It turns out that game theory can help you find remarkable solutions in an unexpected place: the animal kingdom! Although game theory was first developed to understand and explain human behavior—especially economic behavior—it was later applied to the natural world. In the 1960s, the famous British biologist John Maynard Smith teamed up with the reclusive genius George Price to apply game theory to the behavior of all sorts of creatures, from bacteria to orangutans.

Human parents can learn a lot by looking at the strategies that Nature—reflecting millions of years of evolution—has devised to ensure honesty. If it's good enough for fireflies or birds, it might just work for you.

As many game theorists have discovered, lying abounds in the natural kingdom. Take mimicry, for example: it's just a fancy kind of lying. Kevin's favorite story of mimicry involves two species of fireflies, a big one called *Photuris* and a small one called *Photinus*. The smaller firefly blinks its light in order to find mates of its own species: The male blinks in a specific pattern to find a female, who then responds with a pattern of flashes of her own. But occasionally, females from the larger species will blink in such a way as to make themselves look like females from the small species. Poor, unsuspecting males from the smaller species will fly right over to *Photuris*, thinking they've got a chance for a date. The larger firefly gets dinner, but not because the little guy bought it for her. Once she's lured him over, the bigger female eats *Photinus*! Lying at its worst.

While it's pretty surprising *how* these fireflies deceive one another, it's not that surprising that they do it. After all, they are different species competing with one another for resources. Competition and deception go hand in hand. Suppose Jimmy and Laura are playing for the same basketball team. Their interests are the same: They both want to win. When they have the same interests, they never have a reason to deceive each other—what could Laura possibly gain by lying to Jimmy about how he should play? On the other hand, if they're competing in some way, that might motivate deception. If Jimmy and Laura are playing a practice game to see which one the coach selects and which one gets cut, deception is a big possibility. Jimmy might tell Laura to run left when she should run right so that Laura looks foolish and Jimmy gets chosen to play in the big game.

Early on, biologists didn't think that honesty *within* a species would be a problem, because they thought evolution would lead everyone to do what's "best for the species." You still occasionally hear that phrase today. Evolutionary explanations for all sorts of behavior, from fighting to mating, were founded on this idea that individuals evolved to help out other members of their own species. But slowly biologists started to realize that individual animals, plants, and even bacteria compete with others of the same species—each one attempting to produce more offspring than the others. (Richard Dawkins was one of the first to thoroughly explore this idea, in his book *The Selfish Gene*.)

In an early example of game-theoretic reasoning, the American biologist Robert Trivers showed that competition within species even extends all the way to parents and their children. Trivers's work suggested that evolution should lead offspring to have *regular* conflicts with Mom and Dad—over how much time, energy, and resources the parents should invest in their children.

Like so many phenomena in nature, this has its roots in sex. In nature there are two ways to reproduce: asexually—a single individual making an exact copy of itself—or sexually—two individuals coming together to produce a new organism that's a hybrid of the two of them, so the children get some of their genes from each parent. We humans receive half of our genes from Mom and half from Dad. Each time two people have a kid, Nature flips a coin to figure out which of Dad's genes and which of Mom's genes are passed on. Because of this, two siblings aren't genetically identical—they get different assortments of their parents' genes. (Unless, of course, the siblings are identical twins.) Jimmy shares half of his genes with his mother, half with his father, and (on average) half with his sister Laura. Because Jimmy isn't identical to his sister, sometimes Jimmy benefits by taking resources, like food or shelter, from Laura. Parents, on the other hand, have no reason to prefer the big sister to the little brother. So while (genetically speaking) Mom and Dad have an interest in giving both children equal attention, each child wants to try to steal their parents' attention from the other.

As you've probably already discovered, conflicts between siblings quickly ensnare parents. Jimmy might attempt to manipulate his parents into giving him more attention than his sister, but Jimmy's parents don't want to succumb to his stratagems. In the animal kingdom this becomes even more complicated because many offspring don't depend on their parents for very long. Children might want to hamper Mom and Dad's ability to reproduce in the future, while Mom and Dad want to save some of their energy for next year's kids.

That's all great in theory, but does this type of parent-offspring conflict really happen in nature? After Trivers's seminal research, biologists went out looking, and they've found many examples of parents and children coming into conflict. Parents withold resources or actively take them from offspring, and siblings defy parental evenhandedness by squabbling for a bigger share.

Consider the white-fronted bee-eater, a beautifully colored African bird: Bee-eater fathers harass their offspring in an attempt to prevent them from breeding, so that instead they will stay home and help take care of their younger siblings. Surprisingly, this harassment sometimes works. Among chimpanzees, a mother often prefers to have a second child more quickly than her older child would like. The older one knows that when his mom has another baby, he will no longer be able to nurse—and he really doesn't want that day to come. The British zoologist Caroline Tutin has described how young chimps will often try to interfere with their mother's mating in ways that would make you blush—usually by distracting or harassing their mother's new beau. Unfortunately for the kids (but fortunately for Mom), their provocation rarely succeeds.

If the old theory had been correct—if parents and offspring never came into conflict—honesty wouldn't be a problem. But even in the animal kingdom, parents and children disagree about what they want to happen all the time. These conflicts of interest create the possibility of dishonesty. The most well-studied context for this kind of deception is when offspring communicate their hunger to their parents. Baby birds often share the nest with several siblings, and the mother or father bird will bring back food to the chicks. Each hatchling wants to get its parent to give it more food than its siblings—remember, it only shares *some* of its genes with them. But the parents strive to treat each child equally. Because the offspring often want more than their parents can provide, each baby bird has an incentive to pretend to be hungrier than it really is.

In humans, parent-offspring conflict goes beyond handing out dinner to siblings. You have your own goals beyond your kids (the horror!), and these interests may come into conflict with what the little ones want. You might want to go out with friends, while your daughter wants you to stay home and play Scrabble. Even when you're entirely focused on your children, you might disagree with them about how they should behave. Your daughter is probably only thinking about tomorrow or the next day, while you're thinking more long-term. Middle-schooler Kevin didn't want to do his homework because homework was boring. He didn't realize how valuable completing homework would be later in life. Not only would it help him learn math, history, and literature, but it would also give him the skills to follow through on the boring tasks that are all too familiar in adulthood. Luckily, Kevin's parents knew better. While Kevin didn't want to do his homework because he thought that the costs outweighed the benefits, his parents saw advantages further in the future that Kevin wasn't recognizing.

There's more than just caring about the future. We have the perspective to know that our kids' preferences will change over time, too. Children usually don't think this way. Sixteen-year-old Josh may think he'll love his favorite band forever, and should get its name tattooed on his arm. His parents, on the other hand, know that he probably won't even like that band next *month*, let alone twenty years from now. And that's why parents, and the government, too, prevent kids his age from (legally) getting tattoos.

With all the ways that parents and children can come into conflict, you might think that lying is inevitable—that there is no way to create conditions that would encourage honesty. After all, if you want one thing and your child wants another, then your child potentially has an incentive to deceive you in order to get what he wants. Kevin doesn't want to do his homework, but his parents want to be sure he does. However, only Kevin knows how much homework he has in his backpack. How could his parents possibly find a way to keep him honest?

Just as Nature created circumstances for dishonesty, it has also found ways to ensure honest communication. One way around the incentives for dishonesty involves what biologists call an *index*. In brief, an index is a means of communication that cannot be faked. This method of communication is found in amazing places all over the natural world. Take, for instance, the desert grass spider of the southwestern United States. Because food is scarce in the desert, the locations of webs are chosen very carefully by the spiders. And the spiders can be very aggressive in defending their web sites against interlopers. But if an intruder is larger than the homeowner, the smaller spider will usually run away before the fight even starts. How does the pipsqueak realize he's outmatched? By feeling the vibrations of the web as the other spider approaches—heavier spiders create greater perturbations. This way of communicating can't be faked (at least, not until spiders learn to put rocks in their pockets).

Parents make use of indices all the time. One example should be familiar to many of us: If you want to make sure your child has swallowed his medicine, you might ask him to open his mouth and stick out his tongue. There are other ways to communicate that are tough for your kids to falsify. Maybe you are worried that although she says she put her toys away, your daughter just shoved everything into the closet. Ask her to show you where she put each toy. Or maybe you're worried that your son skipped the movie and spent the ticket money on candy. You could ask him to show you the ticket stub from the theater. Are you worried that your son is exaggerating the price of his new school clothes in order to pocket the difference? Make him show you the receipt from his shopping trip.

When kids really start acting out, some parents get creative. Kevin has a friend who had a habit of climbing out of his bedroom window to go hang out with friends and "cause trouble." After several successful nights, he was eventually caught. His parents woke to find his bed empty; they stayed up until he came home and grounded him immediately. But they worried that after serving his punishment for sneaking out, he might try it again. They didn't want to get up every night to check if he was still there. What could they do? His parents came up with an ingenious plan to ensure that he wouldn't sneak out undetected. Each night, they would go outside and put Scotch tape over his window. The next day they would check to see if the tape was still attached. If it was, no problem. If it wasn't, they knew their son must have sneaked out of the house, and there would be hell to pay. The tape couldn't be faked—it was an index.

Sometimes creating an index isn't easy. It would have been difficult for Kevin's parents to prevent him from lying about homework. Perhaps they could have called all of his teachers every day, but this raises obvious difficulties. Not only is it onerous, but it might have had other long-term effects that Kevin's parents wanted to avoid. It might have created an (admittedly deserved) sense of distrust, and it would not have properly reinforced the importance of being honest.

Before they discovered the tape trick, the parents of the sneaky kid used another time-tested way to encourage honesty: They punished lying. Punishments aren't unique to humans—they're found other places in the natural world, too. Many birds indicate where they are in the social hierarchy by growing specially colored feathers. One example is the male house sparrow, which grows a black "badge" on its throat—the bigger the badge, the more important the bird. Every year the house sparrows molt, which gives the males a chance to grow a new badge to represent their new place in the hierarchy. The Danish biologist Anders Møller manipulated these badges with ink. He would make a bird's badge bigger, suggesting the bird was further up in the pecking order. When he put these "liars" back in with the other birds, they were attacked. Those birds with fake badges were punished for lying. House sparrows were able to detect lying—biologists are still perplexed over how.

In order to punish your kids for lying, you, too, have to be able to detect it—you can't punish what you didn't see. Game theorists call situations where dishonesty can always be detected *perfect monitoring*. In many cases you can see the lie, either right away or before the liar has gotten very far. Kevin has a friend who asked his daughter, Eve, if she had stolen all of the carrots from his dinner plate. Eve realized she couldn't answer, because her mouth was full of carrots, so she simply shook her head "no." It didn't take much investigation for Eve's father to discover that his daughter was a carrot thief. Eve was caught by perfect monitoring.

For your penalties to be effective, you have to bear in mind the conditions for effective punishment that we talked about in chapter 4. The punishment must be *credible*—either because you are willing to carry it out or because you've pre-committed (you can't help but follow through)—and the cost of being punished must outweigh the benefit of lying *from the perspective of the child*. If the severity of the grounding is too light, our troublemaking teen might sneak out anyway, because a night of fun is well worth the cost.

There's another wrinkle. If you want to use punishments or rewards to ensure honesty, your kids must expect the punishment and they must expect to get caught. It's not enough that *you* think you'll catch them, *they* must think it, too! In fact, it doesn't matter whether you actually will catch them—just that they think you will. We'd never suggest that you deceive your kids . . . too often. If Kevin's parents had explained that they would get to see his poor grades, or they would ask his teacher about his apparent lack of homework at the next parent-teacher conference, this might have made the threat of punishment more credible.

Once you've convinced your children that you have eyes in the back of your head, what should you use as a punishment or a reward? Game theory suggests you could use almost anything, so long as the punishment or reward meets the conditions we've set out for you. But one very obvious penalty is well known: Stop believing your child. One of Aesop's fables has so institutionalized this type of punishment that it has generated an idiom: *crying wolf*. You'll remember the fable: A young shepherd entertains himself by yelling "Wolf! Wolf!" to attract the villagers, and then laughs at them when they come. Then one day an actual wolf arrives. When the boy cries for help, no one comes, and the boy's herd is devoured. The wise man of the village explains to the boy, "A liar will not be believed, even when he speaks the truth."

The little boy who cries wolf is punished, but not by being forced to go to bed without any dinner. Instead, he's punished by being ignored. For this to be effective, it must be the case that the cost of being ignored in the future outweighs the benefit of getting away with mischief today. In the fable of the little boy who cried wolf, that is clearly the case. He gained only a little attention today, but lost his herd of sheep later. But he did not understand that the future cost would outweigh the immediate rewards. So this wouldn't affect his behavior—at least, not until it was too late.

Alas, this punishment does not always work with real-life children. The threat of future distrust might fail because the benefit of lying today is too great compared to a distant and uncertain future. Or perhaps the threat of being ignored is not credible—perhaps the child thinks the parents couldn't possibly ignore him. And finally, threats of punishment or promises of rewards work only when we're in a situation of perfect monitoring, when your child knows that her lie will be uncovered. Yet children will often lie believing that they'll never get caught—they may think this is not a case of perfect monitoring. Kevin thought he could get away with his homework ruse, although he now admits he had no plan for how to explain his poor grades when they came.

What should we do without perfect monitoring—when honesty is important but we can't observe lying in action? It might seem that this is the most desperate situation of all; you just hope that your child is honest. Nature has found a few solutions here, too.

The first is the most effective: *Change the game.* You can achieve this in two ways: Reduce the benefit of lying, or improve the payoff for telling the truth. To model this solution, let's look at a reclusive seabird called Leach's storm petrel, which lives on islands in the far northern Atlantic and Pacific Oceans. Like many birds, these petrels go hunting—in this case for fish—and return to the nest to feed their young. Leach's storm petrels don't feed their young very often, regularly waiting three days between meals. (Next time your son tells you he's hungry, tell him about these birds!) Unlike many other birds, Leach's storm petrels have "decided" to completely ignore their young when they beg for food. Instead, they seem to feed them on an almost random schedule. This means that petrel chicks have nothing to gain from lying to their parents about how hungry they are.

Now, we're not suggesting that you completely ignore your kids—at least not yet. But we can take a cue from Leach's storm petrel. Parent birds change the incentives faced by their children so that the kids no longer have incentive to lie. How could you put this strategy to use? When it came to Kevin's homework, his parents cut out any rewards he could get from lying. After a few months of deception, Kevin's parents met with his teacher and discovered that Kevin must have been deceiving them about the amount of homework. They realized they couldn't rely on Kevin to tell them how much work he had for a given evening. So, instead, they required Kevin to devote two hours every evening to schoolwork—even if he said he didn't have any to do. His parents would require him to read his textbooks or do extra math problems. Kevin no longer had an incentive to lie, because lying didn't get him in front of the TV any faster.

Reducing the benefit of lying is one way to encourage honesty. The flip side of the same coin is reducing the cost of telling the truth. Suppose Mom comes home from a long day at work to find a vase missing from the living room. It was a tacky gift from Mom's cousin, and she'd be happy to see it gone, but she wants to be sure it wasn't stolen. She asks Ethan if he knows where it is, and he says no. Mom's terrified—perhaps it really was stolen. (What else might be missing?!) Or maybe Ethan broke it and is fibbing in the hope of avoiding punishment.

Many parents have probably already stumbled on the solution to this situation: Explain to Ethan that he won't get punished if he broke the vase. If he just tells the truth, Mom will be happy. He might not believe Mom for a moment or two, but if she can show him that *she's* being honest, he'll probably be honest himself. Before Mom explained the situation, Ethan was worried that if he told the truth—that he broke the vase—he'd get in trouble. But once he realizes that Mom won't be mad, he doesn't see a reason to deceive her anymore. This solution works because it has reduced the cost for Ethan of telling the truth. Now he can be honest without adverse consequences.

Improving the situation of the honest child can be achieved by other means as well. Parents can always reward honesty, either by reducing punishments or giving rewards. Some parents promise they will punish children less if they admit their wrongdoing right away. Or parents might give direct rewards for honesty, such as extra screen time.

For our last mechanism to help ensure your kids' honesty we can turn once more to the natural world and to one of its most fascinating creatures, the beautiful peacock. The peacock has long been a source of wonder. The male's tail is so extravagant that it has served as a symbol of wealth for thousands of years. People over the millennia have also speculated about the origin of this apparently needless excess. The ancient Greeks believed that the tail was bestowed on the bird by the jealous goddess Hera. Hera had a servant, Argus, who was a monster with one hundred eyes. She set ever-watchful Argus to guard Io, one of her husband Zeus's lovers. But Zeus sent the sneaky messenger god Hermes to kill Argus and set Io free. The grieving and furious Hera took Argus's eyes and placed them on her favorite bird, the peacock, to commemorate him as her loyal servant.

Using mythic stories (even ones full of treachery) as explanations for natural wonders may have fallen out of favor after Darwin's *Origin of Species* was first published in 1859, but the mystery of the peacock's tail has only deepened. If it was not created by the whim of a goddess, why would a bird evolve such an ostentatious tail? Darwin recognized that male peacocks use the tail to attract mates. But why do female peahens care? Why would natural selection favor peahens that prefer males with an otherwise useless and cumbersome ornament?

There are several potential explanations for the functions of the peacock's tail, and no clear definitive evidence in favor of one or another. But let's take a look at one of the more popular theories, known as the *handicap principle*, and think about its applications in the world of kids and game theory.

First suggested by a married team of ecologists, Amotz and Avishag Zahavi, after years of studying animal behavior in the wild, the handicap principle states that the peacock's tail is used to signal the fitness of the male to the female. Here's the idea. In peafowl, females are the "choosy" sex—they try to limit their mating to only a few high-quality males. Males, on the other hand, are interested in . . . well . . . you know. Males can be high-quality mates in all sorts of ways: Perhaps they're stronger, faster, smarter, or healthier than other potential mates. Since many of these good qualities will be passed on to her children, female peahens want to mate with the better males; this will produce better children. As a result, peahens will look around and try to figure out who's a good guy and who's just pretending.

As you might guess, this whole situation creates a strong incentive for dishonesty. A low-quality male wants to deceive a female into believing that he is top-notch. In many species with choosy females, nature has evolved a way around this motivation to lie: Males cannot simply "tell" their prospective mates how fit they are. Instead, they rely upon an elaborate signal that imposes a cost—a handicap—that only high-grade males could afford to pay. This kind of "honest signal" is exactly what the Zahavis thought the peacock's ornate tail is. To produce, carry around, and display that tail, the male expends a lot of energy. What's more, its huge size slows him down and makes it harder to avoid predators. He goes to all this trouble to show the female what a superior mate he would be.

Since the Zahavis developed their theory, a number of biologists have given it mathematical precision in the language of game theory. (The Zahavis aren't big fans of mathematics, but that's okay. Game theorists love it!) These mathematical models have shown that—to a certain extent—the theory holds water. But it only works when the fitness cost to grow and maintain the giant tail is lower for the fit male than for the unfit one. For instance, the fit male might have access to a lot of food, so he has the extra calories to spend on a giant tail. Or maybe he's already really good at escaping predators, so he doesn't worry about the tail slowing him down. On the other hand, the poor male might only have a little food and not have much energy to spare, or maybe he needs every little bit of speed he can muster to keep from being somebody's dinner. If he did try to "go for it"—grow a giant tale in order to lie about his quality—he might well die before he could mate.

Some game theorists (including Kevin) have suggested that the handicap principle might not be the panacea that the Zahavis suggest. Nevertheless, it can provide you with some inspiration. You might find ways to make your child's dishonest communication "costly." A common way to encourage honesty is one that you may have already discovered: Ask for more details. You probably ask questions when you suspect your child is lying, because by asking more questions you'll eventually catch your child in a lie. This is a very good reason to fire away and keep the questions coming. But even if you don't catch your child lying, asking questions can have another benefit: It imposes a cost on your child to make up the story in the first place.

Lying is more difficult than telling the truth because lying is creative—your kid has to use his imagination—while telling the truth is simply recalling the past. And making up a story takes work, especially for young kids who aren't experienced at creating fiction. The more questions you ask your child, the more the child will have to work to create a plausible lie, and the more cost the lie imposes.

If Kevin said to his parents that he had no homework, they could simply ask him for more details: What did he do in geometry class? What about physics? When is the next assignment? How did he do on the last test? If he's telling the truth, the answers to all these questions are easy. But if he's lying, he has to make sure that the answer to each question is consistent with his story. And this takes work. For small lies, the effort of making up a story might not be worth the trouble—even if your child knows he could succeed if he really tried.

Just asking once won't do. You need to develop a consistent reputation as the "questioner" so that your child knows the barrage of questions is on the way. Then, she'll have to spend time preparing—coming up with the elaborate, but plausible, story to tell you. If this lie is just to get out of a few chores, she might not even bother. However you do it, though, the cost must be sufficient to outweigh the benefit of lying. Asking a lot of questions will work for lies that have a small benefit, but maybe not for those whoppers where your child has a lot to gain through deception.

Costs can be imposed in other ways. When Kevin was ten years old, he lived in Germany for a few months while his father was there for work. Kevin hated his school, and to get out of it he would pretend to be sick. Kevin's parents knew the dangers of sending a sick child to school, and so they didn't want to just ignore Kevin's complaints. But, after weeks of a strange illness that never arose on weekends, they started to get suspicious. Using the handicap principle might have helped. If they had imposed a cost on Kevin for staying home, he wouldn't have had a good reason to deceive them. For instance, they might have told Kevin that he would have to stay put and do his homework on Saturday when he missed school during that week. Kevin might have realized that it was only worth it to stay home when he was actually ill.

A good game theorist always knows the limitations of each strategy, and this is especially important for you, the nascent game-theorist parent. The example of Kevin's "illness" in Germany also illustrates two pitfalls of the handicap principle. The first is that imposing a cost on your child when it turns out he's telling the truth might seem unfair, even to you. After all, if Kevin is actually sick, he's still getting punished by having to do homework on Saturday. Undoubtedly, our proposed solution would have been met with complaints of injustice. Like the boy who cried wolf, Kevin brought it on himself by lying and creating a sense of distrust between himself and his parents.

That probably wouldn't have prevented him from complaining, however.

And this illustrates another danger of using the handicap principle: While the cost must be large enough to deter lying, it can't be so large that it ends up discouraging *honesty*. If Kevin's parents had told Kevin that he'd get no vacation if he missed even one day of school, then Kevin might opt to go to school even when he was really sick and should stay home in bed. That might be even worse than tolerating the lying. In some circumstances it might be impossible to find a way to make lying costly enough to stop deception without also discouraging your kid from being honest.

And this leaves us with the last, and most disappointing, way nature deals with dishonesty: Just live with it.

Take the cuckoo, for example. This bird lays its eggs in other birds' nests. The baby cuckoo fools the adopted parents into thinking the cuckoo is one of their own. Apparently cuckoo eggs show up infrequently enough that the unwitting surrogate parents haven't evolved ways of dealing with it. And perhaps there are situations when human parents have to raise a cuckoo, too. If you know ahead of time that the lying is of sufficiently low consequence, that you cannot detect it, and that it will occur infrequently, you might do best by not worrying about it.

Nature has found a number of ways around the problem of dishonest children and can provide you with some important guidance as you strive to keep your kids honest. The strategies have their own benefits and pitfalls, and we recommend that you choose one that is best given your own circumstance.

Our takeaways:

- Remember indices, those ways of communicating that make lying impossible. These can help prevent lying, but they don't teach honesty.

- If lying can be detected after the fact, you can reward honesty or punish lying. But remember our advice from chapter 4—the punishments or rewards must be credible.

- Future distrust (ignoring your child when he cries wolf for real) can *sometimes* be an effective punishment.

- Remove the incentive to lie, or compensate for the cost of honesty.

- Make the communication so costly that lying is not worth it.

6

He Started It!

After repeatedly separating sibs and reminding them for the thousandth time that they should *try to be a little nicer to one another*, many of us begin to think we will never put an end to the fighting. We talk about sibling rivalry, but what we're really concerned about is the incessant squabbling that can turn a happy home into what feels like a battleground. That's not rivalry—it's conflict. Parents have many reasons to decide to have only one child, but we wouldn't be surprised if some made that decision solely out of fear of sibling conflict! That might not be such a concern if the battle is between the tennis champs Serena and Venus Williams, or the football quarterbacks Eli and Peyton Manning. But for most parents, this conflict is in the living room, the bedroom, or anywhere else the kids can find a way to lash out at one another without getting caught. Even though nobody's televising it, parents get a ringside seat, whether they want it or not.

But it doesn't have to be this way. Reducing the number and intensity of these conflicts is possible—if we strike the right bargain with our kids. Game theorists have devised several ways to go about this. They know that some kinds of negotiations are more likely to result in fair outcomes for the kids than others. Some schemes require an authority figure to enforce them. In business and government, the enforcer could be a regulatory agency or a police officer. In baseball, it's the umpire. And in the family, it's a parent. Other game-theory strategies are designed to structure bargaining so that no enforcer is needed, making all of the parties—kids and parents—happier. In other words, with the right incentives, kids can be taught to reach fair agreements on their own.

All of this is a big part of what we really want our kids to do with each other—cooperate! We can enforce cooperation. While that might work, it's tough to do. What's more, we don't want our kids to act out of fear. We don't want to be tyrants. Far better would be to create a situation in which the kids figure out for themselves that cooperation beats conflict—and decide to cooperate without threats from the parental authorities. Kids won't cooperate every time. But if they know they must meet again, they may figure out that cooperating this time could win them better treatment from a sib the next time they meet—which could be five minutes later.

If you despair of ever getting your kids to cooperate, remember that cooperation is part of our biology. It's not limited to adults, nor even to humans. Frans de Waal, the professor of primate behavior at the Yerkes National Primate Research Center at Emory University, whom we met in chapter 3, reports that capuchin monkeys are quite capable of cooperation. He demonstrated that concept with this experiment: He set up treats for pairs of capuchin monkeys on a tray that was too heavy for one monkey to pull out, but that could be pulled out by two who chose to cooperate. Two females, Bias and Sammy, did exactly that, except that Sammy quickly snatched her treat and ate it before Bias had time to fetch hers. The tray bounced back, and Bias could no longer get her treat. "Bias threw a tantrum," de Waal reports. "She screamed her lungs out for half a minute until Sammy approached her pull-bar for the second time, glancing at Bias. She then helped Bias bring in the tray again." This time Sammy wasn't pulling for her own treat; she had already eaten it. She was doing the right thing—returning a favor. "Bias had helped her, so how could she refuse to help Bias?" de Waal writes. He didn't find that surprising, because "the group life of these monkeys has the same mixture of cooperation and competition of our own societies." And—we'll add—the same mixture of cooperation and competition we hope to see in our kids.

There are evolutionary reasons why one sibling might want to ensure the survival of another. Mainly it's because they share many of the same genes. And our biological imperative is to preserve our genes. As we saw in chapter 5, though, that isn't always enough to prevent sibling rivalry. (After all, siblings share on average only half their genes.) Cooperation appears in wider circumstances, too. A few years ago, Paul was at a conference at the University of Florida, which maintains bat houses with a colony consisting of some 300,000 bats. The bats emerge in a great black cloud at dusk to consume an estimated 2.5 billion insects every night. Despite the colony's mammoth appetite, not all bats return at night with full furry bellies. These bats get no help from the others. But not all bats are so cruel to their friends. Vampire bats behave differently. Some of those that return hungry are offered regurgitated blood by other bats. And they are likely to return the favor later, when circumstances are reversed. This makes bats *reciprocally altruistic*, because they cooperate with others who are willing to cooperate with them. Some of this sharing occurs between siblings, but some occurs between bats that groom each other—what you might call friends. Cooperation is built-in. If two bat buddies can learn to cooperate, then a pair of human sibs should be able to do the same.

Evidence that suggests they can indeed cooperate comes from the laboratory of Elizabeth Spelke, a Harvard University psychologist and a pioneer in the study of infant cognition. She notes that human adults prefer to share with three groups of people. The first is close relations. The second is people who have shared with us; we want to reward them for that by being generous in return. And the third is people who have shared with others. We like to reward generosity even if it isn't directed at us (game theorists call this *indirect reciprocity*). The problem is that we don't know how we developed those preferences. Are they encoded in our development? Do they come from our experience of others' generosity? Are they values we learn in our religious education? Or do children learn about sharing and cooperation from their families?

Spelke and a colleague, Kristina Olson, did a series of experiments with kids to address those questions. First, they investigated kids' willingness to share with family members. Twenty boys and girls around four years old were each represented by a doll Spelke and Olson call the protagonist. In the experiment, each child's protagonist doll was given resources to share—plastic bananas and oranges, rubber ducks, candy, and more. The children directed the protagonist doll to distribute these gifts to six other dolls. Two of them were described as the protagonist's sisters, two were described as friends, and two were complete strangers. The kids, even at this young age, gave more of their gifts to siblings than to friends, and more to friends than to strangers.

In a similar experiment, children directed the protagonist to distribute more resources to dolls that had been generous to the protagonist than to dolls that had not—demonstrating reciprocity. (The children were read a script explaining which dolls had been generous to them.) And last, the kids gave more resources to dolls that had been generous to others than to dolls that had not been generous to others—demonstrating indirect reciprocity.

The studies, taken together, "provide evidence that three specific principles governing complex, mature cooperative networks emerge early in childhood," Olson and Spelke concluded. Children clearly do not learn these principles from adult experience or religious or moral instruction. The results could reflect their intrinsic development, or what they are learning from other children. Importantly for us, they could also be learning this from the environment in their families.

This kind of conditional generosity—and the special concern for siblings—is related to a child's interest in cooperation, and to the sense of fairness that we've discussed earlier. "Probably fairness evolved to support cooperation in some way. We don't know exactly how that works yet," says Katherine McAuliffe of Boston College, who studies this question. While she and her colleagues don't yet have all the answers, they have hunches. One idea is that we need to know what we're going to get from cooperating, relative to what we're putting out, she explains. "You want to avoid situations where you're being exploited." The presumption is that we're each trying to maximize our gain in any situation. When we use our finely honed sense of fairness to make these assessments, we are then prepared to cooperate when it makes sense for us.

If you think kids can't manage these kinds of calculations, then ask yourself this question: Are your kids smarter than fish? Smarter than the neon tetras, guppies, and angelfish drifting languidly in the small tank your kids promised they would clean but haven't done so even once? If kids are smarter than fish, then they must be able to cooperate—because fish can do it. "It's my favorite example of cooperation outside of humans," says McAuliffe. She's talking about the work of Redouan Bshary, a behavioral ecologist at the University of Neuchâtel in Switzerland.

Bshary has been interested in fish since he was a child. He got his Ph.D. in ecology by studying not fish but tree-dwelling monkeys in Ivory Coast. There he discovered that different species would collaborate to reduce the risk from predators. His work supported a notion that animals operated a kind of market economy in which they would swap food, say, for protection. Bshary began looking for other animals in which he could test this idea, and he found a prime example in so-called cleaner fish, which eat parasites off the skin of other fish, called clients. (We did say this was a market economy.) The cleaner fish get fed, and the client fish get rid of their parasites: cooperation.

Bshary went on to identify all kinds of other interesting social behaviors in the cleaner fish. Cleaners could cheat by supplementing their parasite meals with nips of mucus from a client fish's skin—which caused the client to jump, and sometimes to flee. He found that when cleaner fish worked in pairs, they were less likely to cheat. They also were less likely to cheat when being watched by other client fish. Apparently they didn't want to get a bad reputation as cheaters. (That's one way to enforce cooperation.) He also showed that when grouper fish and coral trout stumbled across hidden prey, they would use gestures to indicate the presence of the hidden food. The fish were doing something that was thought to be limited to large-brained vertebrates—specifically humans, great apes, and, oddly, ravens.

The point is that cooperation is deeply embedded in the lives of many animals, including humans. If fish can learn on their own to cooperate, our big-brained vertebrate kids should be able to manage it.

Consider the Pick-up Dilemma: putting away the LEGOs, puzzles, costumes, and My Little Pony collection that accumulate on the kids' bedroom floors. (Where did we *get* all this stuff?) It's time to clean up, but neither of your kids will budge. You promise them that if they clean up you'll take everyone out for ice cream. Still no luck! Each is waiting for the other to start. Cooperation seems as remote a possibility as kids asking for a handful of kale as an after-school snack.

We suggest taking note of the idea that siblings must negotiate repeatedly over a period of years. What you want is a variation on the Prisoner's Dilemma called the Repeated (or Iterated)

Prisoner's Dilemma. The Prisoner's Dilemma is the game in which two prisoners are separated and given the option of confessing, turning in the other, or saying nothing. If they both say nothing, they both get short sentences for a minor crime. If one confesses and the other doesn't, the confessor goes free and the other gets a longer term. If both confess, they get equal, intermediate sentences. Silence by both would be the best outcome for both—they would get only short sentences. But one of the lasting achievements of game theory was to show that they will both confess. Each wants to guard against the other getting off scot-free.

The Repeated or Iterated Prisoner's Dilemma refers to the situation in which two siblings face the opportunity again and again to keep silent or confess. Now the game becomes more interesting. Your son might choose to tell on your daughter, and she might respond by telling on him, too. But suppose one of them decides not to sacrifice the other. Your son keeps silent instead of blaming your daughter. Maybe the next time, she will offer him the same consideration in return. Why? Because she sees that kindness can be good for both of them. If we can get siblings started along this path, cooperation is likely to increase, with the good behavior of each of the sibs reinforcing that of the other. Admittedly, this is cooperation at the expense of the parents. But we have to start somewhere. And game theorists have proven that once two parties begin to cooperate, that cooperation is indeed likely to increase. They call the strategy Tit for Tat.

We realize that Tit for Tat might not sound like the kind of behavior we want to encourage. It sounds like "you whack me, I whack you." Indeed, dictionaries define "tit for tat" as returning one blow for another—a state of affairs that can quickly escalate into a full-scale brawl. But that's not what game theorists are talking about. They use the term differently from most of us. In game theory, Tit for Tat means that your son, say, begins with a cooperative move. Your daughter can cooperate, or not; but whatever she does, he will do exactly the same thing she does from then on. If she cooperates, he cooperates again. If she refuses to cooperate, or defects, so does he. See the logic? If your son begins with a cooperative move (he picks up a few puzzle pieces) and your daughter does the same (she picks up a few more), they continue in cooperative bliss. If he starts out generously and she rejects the cooperative move (she leaves the room to go and watch TV), he rejects her in turn—and cuts his losses. If she ventures a cooperative move, he returns the favor, and they are back on track. Tit for Tat is the game theorists' version of the Golden Rule—do unto others as you would have them do unto you. But game theorists don't take it on faith; they've proved that it works.

One of the leaders in research on cooperation is Robert Axelrod, a professor of political science at the University of Michigan. Axelrod summarizes much of his work on cooperation in his book *The Evolution of Cooperation*. Axelrod begins by stepping back and asking very general questions. "When should a person cooperate, and when should a person be selfish, in an ongoing interaction with another?" And: "Under what conditions will cooperation emerge in a world of egoists without central authority?" We might paraphrase that one this way: Under what conditions will our children cooperate with us and with one another in a world in which we are not immediately ready to pounce whenever they fail to cooperate?

Some say children will never cooperate in the absence of authority. The great English political philosopher Thomas Hobbes wrote in the seventeenth century that life without an organizing central authority was "solitary, poor, nasty, brutish, and short," a worldview that doesn't leave much room for spontaneous cooperation. Hobbes thought the only way to get cooperation was to install a tyrannical dictator to keep people in line. After a week of silly bickering, parents can see where Hobbes was coming from. Ever since Hobbes, great thinkers have debated whether cooperation could emerge without a central authority (such as parents, or the government) to police it.

Axelrod has a more optimistic view of human nature than Hobbes. People can cooperate even if they are not concerned about the welfare of others or the welfare of their group as a whole, Axelrod writes. He demonstrated this with a computer competition, like a chess tournament, in which game theorists would play the Iterated Prisoner's Dilemma game with one another to see who came out on top. He asked game theorists to submit the strategies they thought would be most successful. Fourteen experts stepped up and submitted entries. The winning strategy turned out to be Tit for Tat, "to my considerable surprise," Axelrod writes.

In strict terms, here's what that meant: A player starts by cooperating in the first round of the game, and from then on does what the other player did last time. Say your son cooperates, and your daughter doesn't. In round two, he doesn't cooperate either. If she cooperates in round three, so does he. The key to cooperation here is that the two parties know they will meet again many times in the same situation. If they played the game only once, neither has an incentive to cooperate. But if they play over and over, your son might figure out that if he cooperates, your daughter may do the same. Remember, cooperation is in their interest if they want to get the ice cream you promised without either one doing more than a fair share of the work. Once they begin to cooperate, they have an incentive to do that every time—and some faith that the other will do the same.

And that's what Axelrod found in his computer game. When he played it with the first fourteen entries, Tit for Tat—submitted by Anatol Rapoport of the University of Toronto—beat other, more complex strategies. It was the most effective at encouraging cooperation in the Prisoner's Dilemma. This was big news. Cooperation was emerging in the absence of any threat from above. Or, as Axelrod put it, "cooperation based solely on reciprocity seemed possible." To be sure of the findings, he ran the computer game again. This time he got sixty-two entries from computer geeks, biologists, physicists, and others. The entries included all kinds of fancy mathematical strategies. Rapoport once again submitted Tit for Tat. Once again, Tit for Tat won.

Now let's get back to the Pick-up Dilemma. When you tell your kids to put their toys away, each one has an incentive to defect, as in the Prisoner's Dilemma. If your son starts picking up stuffed animals and tossing them in the toy box, your daughter's strategy might be to coast—hang back and let her brother clean up the room. The result? The room is cleaned up, they can resume playing, and Mom and Dad are off their backs for a while. The cost to your daughter is nothing, but your son is treated unfairly.

In the Repeated Prisoner's Dilemma, they soon meet again. Somehow the stuffed animals have migrated to the center of the floor, like a herd of wildebeest huddling on the African savannah. Now your son refuses to pick up anything, because he got a shoddy deal the last time.

So this time we institute Tit for Tat, observing and encouraging the kids to adopt this strategy. Your daughter puts away a toy (it's only fair that she start the cleanup this time) and then your son puts one away. They alternate until the room is cleaned up, and they've both contributed equally to the work. You could try alternating days, with your daughter cleaning up one day and your son the next, and so on. The idea is to keep the game going, keep track of whose turn it is, and be scrupulously fair. Each child has the incentive to clean up the room, because doing so will encourage the other one to do the same.

What strategy should each child follow to maximize the likelihood that he or she will get the ice cream with the least amount of work? If one defects, the other must do all the work to get the reward. If they cooperate, each cleans up only half the mess—half of the work.

As with the prisoners in the original Prisoner's Dilemma, the strict pursuit of self-interest by both parties results in—alas—no ice cream. When you leave the room, your son and daughter sit on the couch, each encouraging the other to pick up the toys. If both defect—that is, they remain on the couch—they get no ice cream. Both have opted for the least possible work, with the unhappy consequence that they get no reward.

This situation will occur again and again, every time the room turns into a giant pile of clutter. That's when the children might start to work together. Your son says to your daughter: Remember what happened last time? We didn't get the ice cream. Maybe your daughter grudgingly starts to pick up some Ponies. Maybe your son does the same. If they continue, they will have shared the work equally, and will both get ice cream.

Neither is likely to pick up all of the puzzle pieces alone, because the other one—who did nothing—will get the same reward.

If you think that this won't work with *your* kids, think again. Axelrod points out that it has worked in situations that are far more volatile and dangerous than sibling relationships. One of the most improbable appearances of cooperation appeared in Europe during the bloody trench warfare of World War I. Front-line soldiers, with orders to kill their opponents, devised a kind of Tit for Tat that considerably reduced the bloodshed. In what's referred to as the "live-and-let-live" scenario, soldiers on one side would refrain from shooting to kill—if the other side reciprocated.

The two sides, hunkering down in the trenches for months at a time, were engaged in a Repeated Prisoner's Dilemma game. You might expect that each side would fire as often as possible at the other, which is what they were supposed to do. This was *war*. Yet they knew they would have repeat encounters, so efforts at cooperation might be reciprocated. Once this cooperation was established, it often stuck. "The live-and-let-live system was endemic in trench warfare. It flourished despite the best efforts of senior officers to stop it, despite the passions aroused by combat, despite the military logic of kill or be killed," Axelrod writes. Cooperation can arise even between mortal enemies. The opposing armies were using the Tit for Tat strategy. If one side started shooting, the other side would shoot back—because Tit for Tat says you take the same action your opponent did. If one side hesitates before shooting, however, the other side might hesitate, too. If it does, both sides have experienced a taste of cooperation, in which no soldiers were killed. If each side continues to reflect what the other side did, the shooting stops. If deadly antagonists in wartime can learn to cooperate, then perhaps it's just possible that we can teach our children to cooperate.

Axelrod's computer games made him curious to know "the exact conditions that would be needed to foster coooperation." How, he wanted to know, does cooperation get started? In the trenches stretching across the European battlefronts, somebody had to be the first person not to shoot.

With regard to children, it might seem impossible to get one to make a cooperative move. If you ask your son and daughter to share the task of cleaning up the LEGOs, who goes first? You might try pointing out to both of them that cooperation is the easiest way to get ice cream. And give them the Tit for Tat rule: If your sib cooperates, so should you. To keep a proper accounting of fairness, you might have them take turns picking up a handful of LEGO bricks, until they are all in the box and you can see the floor again.

You could also try a hybrid approach, in which *you* initiate the cooperation. If the kids are locked in a battle in which both defect from the chore at hand, try seeding them with a little cooperation. Your son picks up a handful of LEGOs; then you follow suit. If your daughter refuses, it's your son's turn again. He grabs a handful, and so do you. And suppose you change the stakes. If your son cooperates with you, he gets ice cream—and he does only half of the work. Your daughter, still sulking on the couch, doesn't get any ice cream. She might protest, but she had her chance to cooperate, and she chose not to.

You've now demonstrated how cooperation works. And you've given both kids an incentive to try it. The next time your son picks up a handful of LEGOs, you invite your daughter to do the same. If she does, you have a shot. They know that if they continue in that way, they will both get ice cream. You can leave the room; your work here is done.

The experiment accomplishes several things. It shows the kids that cooperating will get the room cleaned up sooner. It teaches them that they don't need a threat from you to get the work done; they learn that it's easier for each of them if they cooperate. The strategy you've used is to introduce a new player—yourself. You've shown the value of cooperation, and made it more tempting for the kids to join in.

If that doesn't work, you might try a strategy that is the exact opposite: You become the person who refuses to cooperate. It works like this: When the kids refuse to pick up the toys, you say you will step in and pick them up yourself. And neither of the kids will get ice cream. You might find that they quickly realize that they should cooperate to clean up before you do, to get their reward. Your efforts to harshly disrupt the game have prompted your kids to cooperate by giving them a common enemy—you! P. G. Wodehouse's scheming fictional butler Jeeves says of his household, "it was a generally accepted axiom that in times of domestic disagreement, it was necessary only to invite my Aunt Annie for a visit to heal all breaches."

And remember, this situation—a spreading mass on the floor—will be repeated for years to come. The next time, your children might decide to clean up the clutter without your intervention as the spoiler. Cooperation, they might begin to learn, can be easier than defecting.

Axelrod has taken this a step forward with the addition of the concept of generosity. Suppose your son grabs the first handful of LEGO bricks. Your daughter grabs the next. And so it goes until she decides, a few rounds in, that she's tired and doesn't want to continue. According to Tit for Tat, your son should stop, too—the rule is that he does whatever she does. But suppose he believes that she will step up again and help finish the job. Then he should continue to pick up the LEGOs, giving her a chance to change her ways and once again pitch in. This is what Axelrod calls Generous Tit for Tat. Your son allows your daughter to slack off a bit now and then.

Or the opposite could happen. Your daughter refuses to take her turn picking up bricks, and your son stops picking up, too. Your daughter, sorry that she has destroyed the cooperation, decides to pick up another handful of LEGOs, and your son reciprocates. This is what Axelrod calls Contrite Tit for Tat. "A strategy like Generous Tit for Tat is likely to be effective," he writes, as is Contrite Tit for Tat, "because it can correct its own errors and restore mutual cooperation almost immediately."

But this doesn't always work. Spite—the "shady relative" of cooperation (to borrow a phrase from the game theorists Rory Smead and Patrick Forber)—can kill cooperation. We're talking here about the psychological definition of spite—being willing to pay a cost so that someone else has to endure a greater cost.

Spite can destroy sibling cooperation faster than anything, but until recently, little was known about whether children would act out of spite. A few years ago, Katherine McAuliffe took a look at that.

We know that humans are unusual in one respect—they cooperate with complete strangers. Animals generally don't do this. This cooperation probably relates to humans' sense of fairness and aversion to inequity. We've seen that this aversion to inequitable outcomes appears in kids around age four or five. Much of the research that established this was done by McAuliffe and her collaborators and associates across New England. But what McAuliffe wanted to ask now was: *Why* do kids reject inequity—to the point that they are sometimes willing to forgo their own resources to deprive somebody else of a richer reward?

To find out, McAuliffe and company recruited pairs of children, ranging in age from four to nine, and compared them with pairs of adults. The idea was to see who would act out of spite. And there was another possibility. The children were given either fair or unfair shares of candy. Children might reject unfair shares of sweets out of sheer frustration at getting such a bad deal. The researchers designed their experiment to look for both spite and frustration.

In a paper published in 2014, they reported that even young children would act out of spite. That is, kids didn't simply want to reject a bad deal (that's frustration). They wanted to inflict a little pain on the other party. Interestingly, adults did not act out of spite, possibly because they were "more worried than children about not appearing resentful or jealous over candy in front of another adult," McAuliffe and her colleagues wrote. The spite tends to disappear around age eight, when kids begin to feel uncomfortable when they get more candy than another child.

What, then, is the takeaway message for us as parents? First of all, as you know well, your young children will be angry when they get less than their fair share of candy, stickers, or coloring books. The research suggests that four- and five-year-olds aren't capable of any other response. So make sure that the benefits of cooperation are divided equally between the kids. If one stands to gain a lot more from cooperation than the other, spite might just rear its ugly head and destroy the utopia you're trying to create.

Second, while it might seem obvious to us that getting too much candy isn't fair, and that we should give some back, it's not obvious to kids, at least until they are in the fourth or fifth grade. As McAuliffe and her team write, "Our findings suggest that young children show a sophisticated capacity to maintain their competitive standing relative to others, with older children in addition showing concerns about fairness." If they show that concern for fairness, you're on the right track. You're raising kids who are going to be fair and generous—which prepares them to enter that crazy world out there, where these qualities will serve them well.

And third, children can be taught to cooperate. It will take patience, and sometimes resolve, to deny them the ice cream when they haven't cooperated in picking up the LEGOs. But you should hang tough. This can work.

Our takeaways:

- The more often your children interact, the easier it will be for them to cooperate with one another.
- The benefits of repeated cooperative interaction works with friends as well as it does with siblings.
- Be sure to teach your kids how bad behavior today will affect how others treat them in the future.
- Teach your kids the game theorists' version of Tit for Tat as a way to solve cooperative dilemmas.

7

Why Can't *You* Pay for This?

Thomas, a smart kid who is in his elementary school's gifted-and-talented program, is getting reasonably good grades, as he should. Not spectacular, but good enough—if we believe he is really doing his best.

But that's the question: Is Thomas doing his best? Is he working hard and putting in the time required? Is he carefully *thinking* about the notes he jots down for his teacher on *Diary of a Wimpy Kid*? Or is he coasting—writing the first thing that comes to mind and doing just enough to get by? ("I thought this was a good book . . .")

We want to encourage Thomas if he's doing his best, and to give him credit for a job well done. We don't want to lean on him to do better, because he's already doing what he's supposed to do. Pressuring him to get grades that aren't achievable for him is going to make him miserable, make his parents miserable, and might even result in a note from the teacher asking why Thomas doesn't seem to enjoy school anymore. On the other hand, if he's truly slacking off, we want to let him know that that's not a good strategy. We want to turn that around.

What do we do? The answer is that we need a reward system that will give Thomas incentives to work hard—without too much pressure. We want to give him a reason to do the best he can—without giving him the idea that nothing less than an A is acceptable.

Game theorists have developed what they call *principal-agent models* to try to solve problems such as these. In a nutshell, game theorists who've studied this try to create systems of incentives that can overcome the desire to avoid work while not punishing someone doing a good job. Let's see how these models can work for you and your kids.

In most circumstances, parents are the principals in principal-agent models, and kids are the agents. To understand the relationship between principals and agents, think about the agents we see around us all the time. The "agents" in game theory are a lot like travel agents, literary agents, theatrical agents, the agents who represent high-powered athletes, to name a few. And real estate agents—an example many of us are familiar with. They are a good example of the kind of agents that game theorists think about.

Let's say you want to move out of your tiny downtown apartment to a house in the suburbs. You might start your search by turning to the Web and scanning through the listings you find there. That might work, but you will likely find hundreds of listings that could also leave you confused and overwhelmed. If so, you might want to change tactics and engage a real estate agent to help you with your search.

And here's why: A capable real estate agent knows far more than you do about what's available, what's reasonably priced, and which brokers to trust. We could learn all of this ourselves, but unless we're planning to buy and sell houses every day, it would be a huge waste of time. In game-theory terms, we are the principals—and we hire an agent.

But our dealings with the people we explicitly call "agents" represent only a small portion of the agents we encounter every day. We elect politicians, and they become our agents. We hire dentists and doctors to take care of our ills; they, too, are our agents.

A corporate executive's employees are her agents, although she might not think of them that way. The manager can't do all the work that needs to be done, nor does she have the expertise. Let's say her job is to sell a new whiz-bang Silicon Valley widget. She gives the go-ahead, but she doesn't know how to design the prototype, build a mock-up, test it, and set up manufacturing. Nor does she need to know any of that. She hires agents to do that for her. In the same way, parents engage their children—as their agents—to go to school. This might sound weird (or even paradoxical). Our kids are not really aware that they are acting in our interest. If they knew that, they'd probably try to use it to extract more candy from us. But the analogy makes sense when you see how game theorists think about principals and agents.

A game theorist will say that two factors characterize all of these situations: The agent knows more than the principals, and the principals need their agents to do something with that information that the principals can't do themselves. The real estate agent knows more than we do about the market, and the principals need their real estate agents to select the best properties and negotiate for them. Our kids know more than we do about how well they are doing—and how hard they are trying—and we want them to work their hardest. Meeting with teachers can help us determine that, but we still will not know as much as our kids about what's going on in school.

We hope that agents—including our politicians and our children—will report honestly to us. But we have no guarantee of that. The real estate broker might be more interested in closing a deal and earning a commission than in getting us the best price for a new home. The broker could encourage us to "take a peek" at a home priced just outside our comfort zone in order to make just a bit of extra cash. A politician—acting as the agent for his or her constituents—might be more interested in raising money and getting reelected than in serving the voters. (We can hear the vociferous assents from some of you out there.) Doctors and dentists might be more interested in running up their charges than in providing you with the care you really need. We can't be sure, because only a competent dentist or doctor knows what care we need. All we know is what we've read online at WebMD. Agents always know more than the principals they work with.

Economists tell a story about how crazy it can get when one party in an agreement has more information than the other. Two jockeys are debating which of their horses is slower. "My horse is slower," one insists. "No, my horse is slower," responds the other. This continues, and it's hard to know how it can be resolved. If the two race, each can slow down and still claim he is pushing his horse as hard as possible. It works, because only he knows whether the horse is running as fast as it can. He has more information about his own horse than the other jockey does—and the same is true in reverse. The other jockey can also slow down to a crawl, and the issue is never resolved. The game theorists' trick is this: Have each one ride the other's horse. In order to prove that his horse is slowest, each jockey must ride the other's horse as fast as possible. Now the private information doesn't give either jockey an advantage.

Unfortunately, we can't use this kind of solution with our kids. Unlike the jockeys, we can't swap roles with them. And even if we could, it wouldn't give us any information about how hard *they* are working. Thomas—the agent you have "hired" to read and comment on *Diary of a Wimpy Kid* and to learn his times tables as part of his homework—might be thinking about Minecraft when he should be doing his homework.

To understand how this can play out, let's use the example of a manager and one of her employees, an assembly line worker named Jim. He's been with the company for a dozen years, putting the company's trademarked widgets together. As with Thomas, Jim can either work hard—or slack off. The manager needs to know that Jim is doing good work. The manager can't hire another employee or install security cameras to watch Jim, because either option is too expensive—and Jim wouldn't tolerate having somebody constantly looking over his shoulder. To further complicate the picture, the metal stamping machine that Jim uses often breaks down. So even if he works hard, he might not be able to produce a lot of widgets. And sometimes, if the machine is working well, he might produce a lot of widgets even when he's coasting. But in general, if he works harder, he gets more done. Pretty simple, right?

Now here's the difficult part: How should Jim be paid? The

manager wants him to stamp out as many widgets as possible. What kind of incentive can she offer that will best align his interests with hers? Jim, like the real estate agent, has more information, in this case concerning how many widgets he can produce. He knows when he's working hard and when he's not, but his manager doesn't.

Care to take a guess on the incentive? We know what many of you are thinking—pay him based on how many widgets he produces. Sometimes that's unfair. When the machine breaks down, Jim produces fewer widgets and gets paid less. And that's not his fault. On the other hand, sometimes work runs smoothly and he can produce far more than usual, so he gets paid more. On average, he gets paid more if he works harder, although his output can vary from day to day.

This solves the problem of whether Jim is working hard or not. The output tells the story. It's not perfect, but now his incentive is the same as the manager's—produce as much as possible. If you had this idea in mind, you will be glad to know that game theorists agree with you. They have proved that this is the best you can do.

There is a downside to this arrangement, however. This is what's called piecework, and it has a complicated legal and regulatory history. While it often works well for employers, it can push struggling workers to the limit of what they're capable of—or beyond. When Paul was a college student working summers in auto factories in his home town of Detroit, stories were told of exhausted employees who lost fingers or a hand in a stamping machine, because they put in a metal blank and pushed "on" with one hand before they got their other hand out of the way. The machines required the workers to push two buttons simultaneously, so both hands would be out of harm's way when the machine stamped out a part. But workers eager to produce more would tape one button down. This kind of injury is much more common when workers are struggling to produce as much as possible. It's one horrific consequence of establishing piecework incentives. We hear similar horror stories concerning migrant farmworkers, who are often paid on a piecework basis. So this solution is not as simple as we might like.

The exercise we're now engaged in—devising appropriate rewards to encourage workers or our kids to do what we'd like— is what game theorists call *mechanism design*. It's about learning how to deal with somebody who knows more than you do. And it's critical if we want to make the incentives work. Ken Binmore, a British game theorist, has an unusual view of the well-known story that is supposed to illustrate Solomon's wisdom when confronted with two women who both claimed to be the mother of the same baby (an example we discussed in chapter 2). You remember how the story goes: Solomon threatened to cut the baby in half, so each would be satisfied. One woman agreed. The other pleaded with Solomon not to do it and promised she would give up the baby rather than face Solomon's extreme alternative. Solomon, in his wisdom, concluded she was the true mother. He was a smart guy, right?

Binmore doesn't buy it. "The biblical story doesn't support Solomon's proverbial claim to wisdom particularly well," he has written. As we said earlier, Solomon was a natural game theorist, and his strategy worked well in the case of these two women. But he was lucky that the woman falsely claiming to be the mother wasn't a game theorist herself. She could have easily figured out that it was best to say the same thing—that she, too, would give up the baby! Solomon's scheme would not have looked so wise if both women said they would prefer to give up the baby.

The game-theory solution? If both women claimed to be the mother in a legal dispute, Solomon should have established a fine that would be paid by both women if they both claimed to be the mother. Solomon's wisdom would come into play in setting the fine. We assume the real mother would pay everything she has to get the baby back. The false claimant would pay less. Solomon must determine a fine that is more than what the false claimant thinks the baby is worth, but less than what the real mother thinks it's worth. (Those of you who noticed that this is an example of the handicap principle from chapter 5 get extra credit.) Binmore's analysis of the alternative scenarios shows that the real mother will always get the baby. The false mother will abandon her claim, because otherwise she will be forced to pay more than she is willing to.

Now let's get back to Thomas, who isn't working with Jim on the stamping machine but who is stamping ideas in his brain—we hope. We want to reward Thomas for getting good grades. Thomas knows how hard he's working, and we don't. All we see is the output—his grades. We could ask him whether he's working as hard as he can, and he might tell the truth. But if he's thinking strategically and doesn't want to do more work, he might not tell us the truth. Game theorists have a technical term for this fibbing: *cheap talk*.

We know that education doesn't always run smoothly. Some teachers are better than others, and some subjects are harder than others. But on average, if Thomas works more, he will learn more. Following the principles of game theory, we reward Thomas for his individual grades.

But not all reward systems are created equal. Rather than saying, "If you get all As, you can get a new game for your Xbox," we should create a series of incentives. Let's say 1 point for a C, 10 points for a B, and 20 points for an A. With enough points— we decide how many—Thomas earns the game. The idea is that the rewards should get better as the grades get better.

This kind of system must be designed carefully to produce the result we're looking for. Regardless of Thomas's talent for schoolwork, it takes more effort to get an A than a B. So the incentive to earn that A must be big enough to overcome Thomas's aversion to work. If the reward for a B is only a bit smaller than the reward for an A, Thomas might just settle for a B. If the payoff for getting a B is set too low, Thomas might decide that a C will do. With enough Bs and Cs, he might, over time, be able to earn his video game without getting any As. The rewards have to be just right.

There is one other problem with rewarding for grades, and you've probably already guessed what it is: cheating. It's possible to get a higher grade without learning anything extra by plagiarizing an assignment or copying off a test. If the reward you're offering for an A is too great, the incentive could backfire by encouraging cheating. It could also be a problem if Thomas isn't capable of getting an A. Instead of creating an incentive for learning you've created an incentive to cheat—because that becomes Thomas's only way to ever get that Xbox game. That's an unintended consequence, indeed.

Some game theorists we talked to suggested that the problem

of setting the right incentives might be impossible to solve. But others were more optimistic, and they had a suggestion. Parents could employ a strategy analogous to collusion in the marketplace. It works like this: You set up a reward system, and then you give the agent (Thomas) a little wiggle room. Rather than set an A as the standard for kids to meet, we say they have to get at least a B. The smart kid who could get an A might settle for a B. But we're still giving him a reason not to settle for a C. It doesn't guarantee an A, but if Thomas is underperforming, it should give him a nudge. And he might find that he likes learning, and wants to do better in school because he likes it. The unintended consequence this time might be an A!

Before we begin this kind of incentive program, we should recognize that we do know a lot about Thomas. We have indirect ways of eliciting information about his dedication to learning. Suppose, when Thomas is in high school, we take him to visit a few colleges. He finds one that he likes, but it's competitive and he might not get in. Still, he persuades us that he really wants to apply. If he starts getting better grades, then we can conclude that he hadn't been working at his maximum before. Without any words being exchanged, we've learned something. If his work doesn't suddenly improve, we can reasonably conclude he was working as hard as he could all along—because his new college aspirations didn't provoke any change.

So far, we've been talking about a situation in which we, the parents, are the principals, and the kids are our agents. But we can turn that around. If we can imagine that we are "hiring" our kids to do well in school, we can also imagine our kids hiring *us*. What for? Well, to raise them, of course. Just as they know more than we do about what's happening in school, we know more than they do about what they need to grow and develop physically, socially, and emotionally. To make all of that happen, kids need help. They need parents to take care of some very important chores— feeding them, buying clothes, providing shelter, and ferrying them to camp, baseball practice, the chess club, soccer, and all the other activities with which we overschedule our kids. We're not arguing in favor of burying kids in responsibilities, but kids need some of those activities (and want them!), and they can't get them without help from us. They can't drive themselves to sports practice, or sign up for camp and write a check themselves.

And while they might gripe about school, they will need a good educational background to do much of what they will want to do as adults. They are not aware of the bargain they are making, but they are hiring us to help them do well in school, and to prepare them for adulthood. And like the piecework employees in a factory, we sometimes find ourselves pushed to the limit—or beyond—as we provide our children with everything they need to become healthy and successful adults.

So our relationship with our children works two ways simultaneously. In some respects (school) we are the principals and they are the agents. In others (providing food and shelter) they are the principals and we are the agents. In game-theory terms, it's quite a nice arrangement.

This business of preparing our kids for adulthood is a serious obligation. Humans are what biologists call an altricial species, meaning that our offspring take a long, long time to reach the point at which they are mature and can care for themselves.

Most other species have it easier. "No creature in the world (unless, just possibly, a bowhead whale) takes longer to mature than a human child does," writes the anthropologist Sarah Blaffer Hrdy in her book *Mothers and Others*. That might be one reason why human fathers, unlike nearly all other mammals, hang around to help with childrearing. There is a lot of work to be done, for a good number of years. As any single parent will be quick to tell you, it's a tough job for one person alone.

We're tough-minded game theorists, so we don't expect that we, as agents, would provide all of these services to our children for free. How do our kids pay us back for this monumental expenditure of time, money, and the emotional roller coaster of raising children?

They pay us back in all the ways we know about: hugs, kisses, the drawings we put up on the refrigerator door, the painfully out-of-tune band performances at school, those never-ending games of Monopoly, the shared laughs and tears watching *Frozen*, their courage in shaking off a strikeout, their smiles, and their pride in their work. Parenting doesn't always run smoothly. Illnesses, financial difficulties, deaths, unexpected pregnancies, and divorce can complicate this contract. But most of us would agree that parenthood is a contract worth signing. And most of us jump at the opportunity when we think the time is right.

While we are thinking about Thomas in terms of principal-agent models, we should also recognize that he—and we, his parents—face another potential problem. This concerns the issue of *moral hazard*. The term "moral hazard" might sound like some profound character flaw, but it's not that at all; it just happens to be the term that game theorists and economists have settled on. It arises when kids (or corporations) get too much protection. That might sound paradoxical, but too much protection is not good for children or corporations. The concept is often used in the context of business failures, bankruptcy, and insurance, but it applies just as well to possible underperformers such as Thomas.

Here's an example of what game theorists and economists mean by moral hazard. Let's imagine that a family is going to the movies. Linda desperately wants popcorn, but the deal her parents made with her was that she needed to use her own money for treats at the movies. The bad news for Linda is she's spent all of her allowance for the week. "Won't you please buy me some popcorn?" she pleads. Linda's parents know she would love the popcorn, and the movie won't be nearly as much fun without it.

If her parents don't bail out their broke daughter, she will be unhappy and will sulk all the way through the movie. But if they give in just this once and buy popcorn, they've made a serious blunder. What incentive will Linda have to save her allowance if she knows that her parents will still buy her what she wants?

This is moral hazard—an incentive that encourages reckless or undesirable behavior. By giving in, Linda's parents are encouraging her not to save her money, but to be just a little bit reckless with it. Because hey—if she's really stuck, Mom and Dad are there. If her parents say no to the popcorn, they are teaching Linda a virtue—save your money and plan how to spend it.

There is more to this than a giant box of popcorn doused in some oily chemical "butter flavor." It comes up all the time. We're met with countless opportunities to bail our kids out. Last summer, one of Paul's boys was at a playground where another parent was passing out water balloons. Paul's son wanted one, but he wanted Paul to go and get it. Paul, who wants to encourage his kids to speak up more around adults, told his son that he would have to walk over to the parent and ask for it himself. He pleaded for help, and Paul stuck to his position. His son didn't get a water balloon. Had Paul stepped in to help, he would have created moral hazard for his son. Why should he speak up if he knows his father will speak up for him? All that was at stake here was a water balloon, and Paul almost gave in. It took all the conviction he had not to bail his son out.

Moral hazard sometimes involves much higher stakes than movie popcorn or water balloons. Suppose a parent tells a daughter in high school that she must write her college-admissions essay herself. Dad will help with editing, but she has to write it, so that it comes from the heart. The day before the application is due, she's done almost nothing—and she comes to her father for help. If he bails her out of this jam, she experiences moral hazard. If he sticks to the original agreement, she misses the deadline, and maybe misses out on the college she wanted to go to.

That's not the worst example we can think of. Your son in college gets involved with a crowd that uses a lot of drugs, and soon he's using them, too, and getting himself into trouble. If his parents have always bailed him out in the past—bought his popcorn and completed the college admissions essay—he isn't too worried about the consequences of his criminal activity; his parents will do their best to bail him out if he gets caught. This happens all the time, even among adults outside the realm of parents and children.

Here's an example from the business world: Owning insurance makes companies more reckless. They know they are insured, so why take extra time to be certain their equipment is being used as safely as possible? If something happens, they're covered! Critics raised concerns about the bailout of big American banks a few years ago for exactly this reason. Some of the banks responsible for the economic crisis got millions of dollars from the government—and they might well conclude that the government will bail them out if they get in trouble again. That's moral hazard in global terms.

And Thomas, our possibly underperforming student, is also potentially subject to the ills of moral hazard. If we help him too much, or spend too much time talking to his teacher to get him help with his grades, we expose him to moral hazard. Why should Thomas work hard if he knows his parents will step in when he has problems? It's insurance, and because he knows he has it, he relaxes a little bit too much.

Game theorists who have worked on moral hazard have a strong argument to make for not bailing our kids out of trouble. A better idea, they suggest, is to establish firm rules ahead of time. The idea will be familiar to many of us. Game theorists argue that there is a difference between choosing what seems best at the moment and adopting a rule that seems good and sticking to it.

Parents have all experienced a situation where they have adopted a rule—no TV after 8:00 p.m.—and then there comes that special circumstance in which they'll want to bend it *just this once.* Being willing to bend it just this once creates moral hazard. If your kids know that you'll bend the rules, they will put you in situations where you'll feel obliged to bend the rules more often. And then the rule will evaporate.

Imagine that you are in a candy store with your son, and he comes up to you with two candy bars and asks you to buy them. "No," you say. You might not realize it, but you're already at a strong disadvantage with respect to your game-theorist son. "You can have one," you respond, thinking you've done your duty as a parent. Your son smiles, says thanks, and tears open the wrapper. By being willing to compromise, you have placed the boy in moral hazard. He only wanted one candy bar; so for him it wasn't a compromise at all. And because he knew you well enough, from past experience, to know that you would compromise, he asks for two. You have pushed him into making a more extreme demand than he would have made otherwise. If you buy both, the situation deteriorates even more. Next time he might ask for three.

Our solution to this problem of pushing kids to extremes is to establish rules ahead of time, and then don't ever bend them—not even just this once. We discovered the importance of this with punishment: Being consistent with threats teaches kids that you're serious when it comes to repercussions. That makes them less inclined to break the rules—and means you might never actually have to apply the punishment, which is what you'd prefer. The same is true with moral hazard. Be consistent, make sure the kids understand the rules and follow them—and whatever happens, don't bail them out.

. . .

To sum it all up: Principal-agent models can help you understand the relationship between you and your children when you're negotiating rules for homework. And an understanding of moral hazard can help you know when you should refuse to compromise.

Our takeaways:

- Make peace with your lack of information—there is only so much you can do.
- Set up incentives to encourage your kids to do well and work hard, but don't make the incentives too big—or too small.
- Reward small achievements, but give bigger rewards for bigger achievements.
- Watch out for moral hazard, don't bail your kids out, and stick to the rules you established ahead of time.

8

Are You Saying You Don't Believe Me?

When Kevin was eight years old, he desperately wanted a cat. He doesn't even remember what got him interested in the idea. But whatever the reason, he *really* needed one. He asked, and asked, and asked again—but his parents weren't so sure. Perhaps this was just a passing phase. And, what's more, they didn't know if *they* wanted a cat. So they did what all good parents do: They stalled. First, they said the family should wait until they moved into a new house. Then, they had to wait until after the family spent a few months in Germany for his dad's work. But eventually, Kevin's parents ran out of excuses.

They relented, but only if Kevin promised (*promised!*) that he would take care of the cat. He would be responsible for feeding her, for cleaning the litter box, and for letting her out and letting her in (and out and in again). Kevin agreed immediately. After all, what's a little time cleaning up litter?

This story doesn't have a surprise ending. Kevin and his parents go to the animal shelter and pick out a small, and very loud, kitten. The family brings her home, and Kevin takes care of her for a few months. But slowly, bit by bit, his parents start picking up the slack. After a while, they're doing almost all of the work. They did a good job—the cat lived for twenty-one years!—but they surely didn't have Kevin to thank for her longevity.

Kevin wasn't trying to intentionally deceive his parents. He really wanted a cat, and he was willing to take care of it to get one. But he didn't follow through on his promise. Kevin let himself get distracted. He'd let the litter go a little too long. He'd forget to feed the cat sometimes. And the cat would make it very clear to Kevin's parents that something needed to be done about the situation. His parents loved the cat, too, of course, and they weren't about to take it back to the pound.

Of course, agreements like this aren't only between parents and children. Siblings strike deals with one another, too. Kevin's friend Rachel has a younger sister who would constantly invade her privacy and barge into her room when they were in grade school. Rachel's sister would have a good ol' time rearranging the stuffed animals in a way more to her liking, trying on Rachel's clothes, and moving the furniture around. As anyone with younger siblings could guess, Rachel hated this. One day she made her little sister promise (*promise!*) to never come into her room again. Rachel's little sister promised, but Rachel didn't think about how to make that agreement stick. Later that same day, she went out, and returned home to find her little sister happily hanging out in her room.

Parents make (and sometimes break) promises with each

other, too. It seems like every parent of a newborn has a story like this one: Mom and Dad are awakened by crying (again). They debate about which one will handle putting the little one back to bed (again). Dad promises (again) that if Mom handles it tonight, he'll do it tomorrow night. But, come tomorrow night, Dad wants to renegotiate the agreement (again).

While the word "contract" has a bit of a legalistic ring to it, all three of these promises are just that: contracts. One party agrees to do X in exchange for another party doing Y. All the world's economies run on contracts, and game theorists have spent a lot of time thinking about them.

In the language of game theory, all these agreements failed because the parties to the contract did not create a means of enforcement. Although Kevin had every reason to agree to the plan beforehand, when it came time for him to carry out his part and get the Friskies cans out of the pantry, he wanted to play instead. There was nothing to discourage him from shirking his responsibilities; Kevin knew the cat could rely on his parents. Rachel's sister didn't keep her promise because she didn't see a consequence to ignoring her big sister's demands. And Dad will try to renegotiate his promises because he doesn't realize that there are consequences for reneging on baby duties.

We are all familiar with contracts that require outside enforcement; our everyday contracts are like this. When you sign the paperwork to buy a house, you know that you'll get the house (assuming you put up the down payment, of course). If the seller tries to back out, the government will step in and help you enforce the contract. As anyone who's bought a house will tell you, there is a lot of gray area—how clean will the house be when you move in? What will be left and what will be taken?

But the basics of the contract are adhered to, and if they aren't, the seller (usually) will be punished.

Kids constantly rely on outside punishment to enforce their contracts with their friends and siblings, too. That's why playtime always features that familiar refrain: "Mooooooommmmm!" But wouldn't it be better if kids could find a way to make contracts that don't require appeal to the Supreme Court of the House: Mom or Dad?

Game theorists have been interested for years in contracts that don't require this "outside" enforcement from real estate brokers, parents, or other VIPs. Scholars call these *self-enforcing,* and you can put these to use in your household. Self-enforcing promises feature prominently in many facets of life. There is very little external enforcement available in international relations, for example. Treaties and trade agreements long predated the creation of the UN or WTO. Despite their lack of external enforcement, agreements for an end to war, for free trade, for nonaggression, and for mutual defense have been around for more than four thousand years. In many cases they feature mechanisms to enforce the agreements, such as embargoes or the withholding of aid, that don't require an appeal to a third party.

Criminal enterprises can't turn to the police to enforce their contracts. Despite the lack of external enforcement, sometimes criminal enterprises succeed in making and keeping agreements between their members and between rival gangs. In the United States, criminal enterprises have been working together for well over a century. And when they fail, sometimes spectacularly, they often do so because the agreements cease to enforce themselves.

Clearly, we humans cooperated before there even was a government at all. There remains a lot of speculation about how exactly we, or our primate ancestors, came to live in groups, but however it happened, it could not have involved an appeal to a government to enforce our agreements. Governments could only have developed after we agreed to come together and form groups in the first place.

While most everyday business transactions rely on external enforcement, governments, criminals, early humans, and you—parents—can benefit from relying on agreements that enforce themselves. Governments have no one to turn to when the other side reneges on their treaty obligations. Criminals, for obvious reasons, don't want to turn to the government when another criminal fails to follow through. Early humans had no government to help out. And you don't want to have to step in to settle every dispute between your children, nor do you want to call Grandma to come in and enforce your agreements for you.

Self-enforcing agreements lie at the heart of one of the most famous ideas in game theory: the Nash equilibrium. The idea is named after its inventor, John Nash, whom we met in the introduction. Nash developed the concept of an *equilibrium* in games, which over several years came to bear his name. Nash strove to predict the outcome of any interaction where people depend on one another.

Nash's idea is one that's best illustrated through examples; here's a simple one. Suppose that Billy and Suzie agree to a plan: Billy will give Suzie his baseball glove in exchange for Suzie's boxed set of the Hunger Games trilogy. To ensure that nobody tries to run off with both prizes, they agree to make the exchange simultaneously.

This agreement is a Nash equilibrium if Billy is willing to follow through on the agreement, assuming that Suzie will, too; and if Suzie will follow through on the agreement, given that Billy will also. That's a long sentence, but the idea is simple. If Billy wants Suzie's books more than his own baseball glove, he's happy to make the trade. And if Suzie wants Billy's baseball glove more than her books, she's happy to make the trade, too. So the trade is a Nash equilibrium.

To really get a grip on Nash's insight, we can also look at one of the idea's early predecessors. More than one hundred years before Nash, the French philosopher and mathematician Antoine Augustin Cournot looked at the possibility of "equilibrium" agreements in a very particular situation. Cournot considered two businesses that are both producing the same product—think two farms of the same size that are both producing corn. Cournot asked: How might these farms decide how much corn to produce? This isn't a simple decision, because how much you want to produce depends on what the price of corn will be at harvest time. And the price depends on how much corn the other farm produces, too.

Cournot imagined that each farm tries to maximize its profits in light of what the other is doing. Through a process of slow adjustment, the farms would come to find a happy compromise where neither wanted to produce more or less than what they actually produce. If they produced more the price would fall, and they would lose money. If they produced less, they would have less to sell and they would again lose money. So, Cournot showed, there is an amount of corn that is a happy medium between the urge to produce more and the urge to keep the price high.

What Cournot argued is that if each farm produces this special amount of corn, each farm is doing the best it can, *given what the other is doing*. This is what makes it an "equilibrium" for this situation—neither farm has any reason to change how much they are producing. You could also think about this as a self-enforcing agreement, because neither farm would want to increase or decrease its production under the assumption that the other farm keeps producing the same amount. Imagine that the two farmers got together to agree on an amount to produce. If they agreed to produce the "equilibrium amount" they wouldn't require a way to enforce the agreement—both sides would want to keep the agreement. But if they tried to agree to produce less in order to keep the price high, they would need some way to enforce it. One farmer would want to (secretly) plant more corn.

Cournot showed that these equilibria exist for many different situations that are analogous to the two farms—where there are two producers of an economic good who compete in a marketplace. But Cournot didn't go beyond this to consider more actors or other situations beyond this one type of economic interaction. Von Neumann and Morgenstern showed that this kind of equilibrium exists for any two people playing a zero-sum game. (We talked about those way back in chapter 1.) Just as the name suggests, a zero-sum game is a game in which one person's gain is another's loss. In the end, it was John Nash who saw that all of this could apply to any type of social exchange whatsoever, from zero-sum to completely cooperative and from two people to two billion people. Nash showed that the equilibrium explained by Cournot, von Neumann, and Morgenstern exists everywhere. And this is what resulted in a Nobel Prize for Nash.

If you're a fan of the movie version of *A Beautiful Mind*, you may be distressed to hear that the movie's example of a Nash equilibrium presents a misleading explanation of Nash's insight. The fictionalized John Nash is sitting in a bar with his (all-male) graduate school friends. In walks a beautiful woman and several of her (less stunning) female friends. All the men's jaws drop at the sight of the beautiful woman. And then, as the movie goes, fictional John has an insight. If they all try to woo the beautiful woman, they will interfere with each other and no one will get anywhere. However, if they all decide to ignore the beautiful woman, and instead shower affection on her friends, they each stand a chance.

It's a true insight as far as it goes, but here's the problem. Suppose each friend agrees with fictional Nash and says, "Okay, I'll ignore the beautiful woman on the condition that you do, too." Is that a self-enforcing agreement? Nope. Why not? Well, let's suppose that each of John's friends keeps to his part of the agreement—each of them ignores the beautiful woman. Nash needs to ask, should I keep ignoring the beautiful woman now that my friends are ignoring her? The answer is probably no, since he doesn't run the risk of interference from his friends. So the agreement to ignore the beautiful woman is not a Nash equilibrium. We're sorry to conclude that the filmmakers were not game theorists. On the other hand, we're not very good at making movies.

Just because the agreement in the movie isn't self-enforcing, it doesn't mean that there aren't any self-enforcing agreements that might have been used instead. For example, the friends could choose ahead of time which one gets a chance to woo the beautiful woman. Once they've agreed on that, each friend now has an incentive to keep the agreement. If one switches from entertaining the less attractive woman to the beautiful one, he doesn't stand a chance because his friend is already there.

Kevin's agreement with his parents to take care of the cat was not a Nash equilibrium, because Kevin did not have to act until after his parents did. They got the cat first, and only then could they know whether Kevin would follow through. So, once Kevin's parents had already done their part—they got the cat—Kevin had no incentive to keep his part of the deal. Similarly, Rachel's sister had no reason to stick to her promise and stay out of Rachel's room once Rachel left. And lazy Dad had no reason (or so he thought) not to try renegotiating his promise the night after Mom took care of the baby. None of these agreements were Nash equilibria, and so they failed.

It's not just Kevin's promise to his parents; any agreement involving parents and children will work better when it's self-enforcing—when it's a Nash equilibrium. In order to figure out if an agreement is a Nash equilibrium, you start by thinking about whether each party has an incentive to keep to the agreement if the other does.

Here's one more: Suppose that Hannah's parents make an agreement with Hannah that they'll let her spend the night at her friend's house if she picks up her room. Mom and Dad would be happy for an evening to themselves, so the agreement is self-enforcing from their perspective. Is it for Hannah? How much does she like her friend, and how much does she hate cleaning her room? If she likes spending time with her friend more than she hates putting her toys away, the agreement is self-enforcing. But if the joys of staying up late telling ghost stories aren't enough, then she won't keep her part of the bargain.

This is what's so tricky for parents and siblings alike: They must think about the terms of a contract *from the perspective of each party.* Kevin's parents recognized that Kevin *ought* to keep his part of the litter box deal. After all, he would need his parents' trust time and time again in the future. But it turned out Kevin didn't think that way. Kevin didn't think about the future that much, and so he ignored the long-term costs of breaking his promise. So while his parents thought the agreement would be self-enforcing, it wasn't. Rachel made her sister promise not to come into the room. But she didn't think about what reason her sister had to keep her promise. She didn't think about it from her sister's perspective. And finally, hardworking Mom made the fateful mistake of thinking that her husband recognized the long-term costs of a sleep-deprived partner.

In all of these stories, one of the parties made a rookie mistake—failing to think about the agreement from the point of view of the other person. Had they thought like game theorists and considered whether the other party would want to follow the agreement, they might have seen their error. It's not enough that Mom and Dad think that Kevin *should* follow through; they need to think about whether he *wants* to.

It's all fine and good to point out others' mistakes, but you want to know how to design agreements that will actually enforce themselves. So, now we'll turn to a few types of self-enforcing agreements that game theorists have discovered over the years—and which could be adapted to families.

The first of these is called a *convention*, an agreement in which it is in the short-term interest of all parties to stick to their promises. Think of two people driving toward each other on a narrow road. In the United States, each will move to the right in order to pass by one another unscathed. They do this not so much because of the law, but because each of them expects the other to do the same. No one has an incentive to switch unless, for some crazy reason, they think the other will as well. That's what makes conventions great: It's in everyone's interest to keep following them.

Long before anyone developed mathematical theories of social behavior, the Scottish philosopher David Hume argued that conventions are a central part of human life. Hume's classic example of a convention is two people out on a lake in a rowboat. Even the most type A personalities rarely make explicit agreements about how they will row. They just fall in rhythm with one another. And, once they're in a rhythm, they just keep rowing at the same rate. Why? Because they both want the boat to move toward their destination. If one were to speed up or slow down, they would end up going nowhere fast. So they stay in sync. In modern terminology, their rate of rowing is a Nash equilibrium.

Hume used simple conventions—like the rowboat—as analogies for many rules of society. You fall into these rules almost by accident, and then you want to keep following them—so long as everyone else does, too. Hume gave us many examples, including the right to private property, the use of money, and human language.

Hume posited that these rules, rather than emerging by explicit agreement, might have emerged slowly over time as people discovered their usefulness. Adam Smith, the father of economics, used a similar idea to argue that markets lead to socially good outcomes. Adam Smith posited the "invisible hand" that guided society toward better and better outcomes as a result of many small choices by the society's members. It's probably no accident that there is a great deal of similarity between Hume's theory about the emergence of conventions through accumulated little changes in the behavior of many people, and Smith's idea about how individuals' everyday economic choices steer whole economies. Hume and Smith were friends. Although Hume was writing long before the invention of game theory, modern philosophers, such as Brian Skyrms at the University of California, Irvine, are finding they can rephrase Hume's insights in the contemporary language of game theory.

Conventions, like rowing a boat or driving on the right side of the road, are one example of self-enforcing agreements. We've never heard of rowboat police who are charged with keeping everyone in rhythm. And most people don't need the government to tell them to drive on the right—the fear of a collision is enough to keep them inside the lines. Kids can establish conventions, too. They might agree on who gets to be which prince, princess, or action hero. Twins might agree to dress differently so that people can tell them apart (or always dress the same so no one can). Parents and kids can have bedtime or dinnertime rituals that they share and mutually enjoy. Or a mom and her daughter might have special names for activities that only they know of. (Conventions aren't all great news for parents: Siblings might agree to keep quiet about secret bad deeds so no one gets into trouble.)

While John Nash showed that every social interaction has *some* kind of self-enforcing agreement, sometimes that agreement is pretty bad for everybody. Kids might adopt a convention of mutual distrust—"I won't trust him and he won't trust me." We gave you another example of a bad social convention in chapter 6. Remember the story of two kids cleaning up LEGOs? Mom told the kids that if the room was cleaned soon, she would take them both out for ice cream. It goes without saying that both kids wanted ice cream, but each one wanted to leave the cleaning up to the other. And by each waiting for the other one to clean the room, neither of them got any ice cream.

The problem with this story is that agreeing to pick up the room doesn't work as a self-enforcing agreement. Even if they agree to work together and clean up, each has an incentive to do a little bit less, leaving the hard work to the other one. But recall that there's a solution! When the interaction is repeated over and over, we have the possibility that both children might agree to cooperate (if the other cooperates, too). This works if the future benefits are big enough to outweigh the short-term temptation to break the agreement.

Most promises that kids make to each other and to their parents involve some trade-offs between short-term costs and long-term benefits. Taking turns works like this. By letting Alyssa play the video game now, Kyle is giving up the possibility that he might enjoy the game now in exchange for more time on it later.

We usually rely on future benefits to help enforce agreements. Kevin's parents thought that Kevin would follow through on his promise to let the cat out because he knew that his parents' distrust in the future would be bad for him. Rachel thought her little sister would keep her promise to stay out of Rachel's room because she wouldn't want Rachel to retaliate. And our fictional mother of a newborn thought her sleepy husband would keep his promise to get up tomorrow night because maybe, just maybe, he'd like to stay married.

In all three stories future benefits didn't work. Why didn't Kevin appreciate that by ditching his cat duties, he would lose credibility with his parents? Experienced parents can probably already give you the answer—kids just don't think this way. And that's no surprise, because sometimes adults don't think about the future this way, either! Adults and children often treat future benefits as less valuable than more-immediate gains. That's why Kevin was willing to go outside and play instead of taking care of the cat, even though in the long term it would lead to a bad outcome—his parents' distrust. And that's why Dad is willing to put off taking care of the newborn again and again. He can sleep in now, even if it will cause problems for his marriage in the long term.

Game theorists are very interested in how both children and adults compare short-term costs and long-term gains. And how does this evaluation of time influence when promises will be kept and when they will be abandoned? There have been countless experiments on how adults compare long-term gains with short-term losses, because these trade-offs affect many important life decisions. Saving for retirement is a clear example.

People are taking money they could use today and putting it away for later. If their investments grow, they will get more when they retire than they are giving up today. So why doesn't everyone put away as much as possible? After all, by investing wisely, they end up with more lifetime income than if they use the money now.

This question was first addressed by a relatively unknown thinker, John Rae, in a book published in 1834. Rae, a Scot turned Canadian, criticized Adam Smith, who was very popular at the time. This put Rae on the outs with many thinkers, and as a result he was largely ignored until the philosopher John Stuart Mill publicized his work many years later. Like Adam Smith, Rae wanted to understand why some countries had significant investment in industry and became wealthy while other countries did not. Rae argued that the way a country's citizens thought about current and future consumption was one of the central differences. Whether or not Rae was right about the cause of global inequality, he began the study of how people think about and value the future.

Rae argued that if we knew that we would live forever, and if we were perfectly rational, then we would always invest our money for the future. But, alas, we don't live forever. And, more important, we don't know how long we'll live. So giving up a treat today for more later may be a waste if we don't live long enough to get the payoff.

Rae's explanation also shows why many agreements fail to be self-enforcing. For Kevin, the future was too uncertain. He didn't really understand how bad it was going to be if his parents couldn't trust him—that seemed so nebulous and distant.

Rachel's sister couldn't really appreciate how bad her sister's revenge would be. Dad failed to see how uncomfortable his wife was going to make him tomorrow.

Studying exactly how people make decisions to delay benefits today for greater benefits tomorrow is one of the most active areas of study in behavioral economics and game theory. Its conclusions have been applied to a diverse range of problems, from tax policy to drug addiction. While most of the studies focus on adults, some have been conducted on children.

The well-known Marshmallow Test, for example, offers a particularly clear glimpse into thinking about how children deal with future rewards. The experiment was first conducted by psychologist Walter Mischel and his colleagues in the 1970s, and it has been repeated many times since. There are many small variations on the experiment, but the basic idea is this: Put a young child in a room with some yummy treat (often a marshmallow, as the name suggests). Tell the child, "I'm going to leave, but I'm going to come back in a few minutes. You're free to eat the marshmallow if you like. But if you wait, I'll give you that marshmallow plus another one when I return."

In many versions of the experiment the psychologists videotape the kids while the experimenter is away. The results are both adorable and hilarious and can be found all over the Internet. Kids try to distract themselves, try to talk themselves out of eating the marshmallow, and try to get around the rules by licking or nibbling at the marshmallow. Some are able to wait while others do not.

The study received an enormous amount of press because, in a follow-up study many years later, Mischel found that the ability to delay eating the marshmallow predicted success in a number of other areas of life—thus setting in motion the terror of millions of parents that their children would fail at the Marshmallow Test.

Mischel and some psychologists view this test as one of self-control. The kids understand intellectually that they want two marshmallows, but they are drawn to the yummy one directly in front of them. The battle between these two processes—the emotional one that draws us to the delicious marshmallow now and the rational one that tells us to wait—is familiar to many of us. Those of us who've tried to diet, reduce our spending, or exercise more can certainly relate to the internal struggle exemplified by the kids' valiant efforts to resist that delicious marshmallow.

How would Mischel describe our three stories of failed agreements? For Mischel, Kevin might have understood how bad it would be if his parents didn't trust him. But when it came time to come in from playing in the yard, he just couldn't overcome the appeal of "one more minute." Rachel's sister knew she shouldn't play in Rachel's room, but the draw of dressing up in Rachel's clothes was just too strong. And, finally, lazy Dad knew he needed to get up and take care of his daughter, but he just couldn't bring himself to get out from underneath the warm covers.

In order to make these agreements self-enforcing, we need to teach kids (and ourselves) self-control. In his book *The Marshmallow Test*, Mischel describes how children can be taught strategies to help them avoid temptation. Mischel explores two ways of thinking about rewards. The one he calls "hot" is the emotional connection to the reward. Thinking about how delicious that marshmallow must be and how great it's going to taste is an example of hot thinking. The other form of thinking, which he calls (you guessed it) "cold," is more rational: The child says "I want more marshmallows, so I'll wait."

Mischel says that the children who "succeeded" at his test were those who were able to circumvent the hot style of thinking. They would distract themselves with games. They would look away, or imagine that the marshmallow was not a marshmallow but a fluffy cloud. These strategies develop with age—Mischel says they tend to emerge between four and five—but they can also be taught. Helping children see how to control their hot thinking and use cool thinking instead will help them to resist life's constant temptations, from fatty foods to spending money they don't have.

And cool thinking will help to make more agreements self-enforcing. If Kevin had been better at cooling down his thinking, perhaps he might have put more stock in his parents' future distrust. Rachel's sister could have used cool thinking to stay out of big sis's room. And cool thinking would have gotten Dad out of bed.

Mischel's way of looking at this test is shared by many psychologists, but some view the results of this experiment differently. They see the experiment as a test of how children value current rewards as compared to an uncertain and distant future. Those who eat the marshmallow immediately simply want the yummy taste now more than an uncertain amount of yumminess later. This way of looking at the experiment coincides with how game theorists tend to think about how adults weigh future losses against short-term gains.

In a variation on the marshmallow experiment, the psychologist Celeste Kidd and her colleagues found that children who had promises deliberately broken by the psychologist immediately prior to the experiment were more likely to eat the marshmallow. She argues that children who eat the single marshmallow may do so not because they lack self-control but because they are more uncertain about the honesty of the person who promises more marshmallows in the future. If they are sufficiently unsure about the benefit of waiting, then the rational thing to do is eat the marshmallow that's there right now. Kids who take the marshmallow and those who wait may be equally "cold" in their thinking—they just have different expectations about the future.

One of Kevin's game theorist friends, Cailin, sees that her twin toddlers treat the future as distant and uncertain. One of Cailin's daughters would gladly agree to turn off the TV in five minutes in exchange for a piece of candy now. But when it actually comes time to turn off the TV, it's the greatest injustice in the world. (Sound familiar?) On the other hand, Cailin's daughter would have a hard time forgoing even the smallest benefit today in exchange for a wonderful gain tomorrow. Put in terms of Kidd's theory, Cailin's young daughters disregard what might happen in that distant world.

Under Kidd's theory, Kevin didn't feed the cat because he was unsure what the value of his parents' distrust would be. Perhaps he would never need to make another agreement with his parents. (If he thought that, he was badly mistaken.) Rachel's little sister went back into Rachel's room because she didn't think that Rachel's future trust was worth worrying about. Poor takeaways:

Help your kids keep their promises by making self-enforcing agreements that the kids will *want* to keep.

Conventions—promises that are in everyone's short-term interests—are the most powerful self-enforcing agreements. Use them when you can.

Being consistent with your kids and teaching them to employ "cool thinking" will help them to put off short-term temptations and keep their promises.

Teaching empathy will help kids see how broken promises hurt other people and help them to develop other-regarding preferences.

Dad severely underestimated how often his situation would come up.

So how can you use Kidd's theory to get your kids to care more about the future? Being consistent will help. As children come to expect you to behave in a steady way, they will become confident that those future benefits will really materialize. You might practice delaying rewards at home as well. Ask your child to wait a little bit before getting dessert, and if they do they will get a little more. Or have your kid wait an extra day in order to get a slightly bigger allowance. They might come to expect future rewards to materialize if they see promises come to fruition over and over again.

As your children develop the expectation that future gains will materialize, you will be able to make more self-enforcing agreements. Now your kids will appreciate that it's worth skipping a short-term benefit in order to secure a much larger long-term gain. Once that happens, you can use future rewards to encourage your children to keep their agreements today. Kevin's parents can explain to Kevin that if he keeps his promise to let the cat out every morning, they will be more willing to let him take on responsibilities later. If he's good with the cat, he might be allowed to go out and play by himself next year. Rachel could have explained to her little sister that if she stays out of Rachel's room, then maybe Rachel will let her play with her toys later. And Mom can explain to Dad that if he doesn't keep his promises, in a few months he might not have a wife to make promises to. In all three cases, the reliable prospect of future rewards or punishments makes the agreement self-enforcing.

A lot of ink has been spilled over whether Mischel's or Kidd's way of looking at the Marshmallow Test is correct. As parents, it probably wouldn't hurt to approach your children using both theories. Kidd argues that both explanations are probably involved when your kids decide whether it's worth delaying gratification now for a better payoff in the future. No matter which theory is right, both argue that you can help your kids to forgo short-term excitement for a significant gain in the future.

There is one last trick for making agreements self-enforcing: Teach your children to develop empathy. When children start to care about what happens to other people, beyond themselves, agreements can switch from being unenforceable to self-enforcing. Rachel's sister didn't stay out of Rachel's room, in part, because she failed to appreciate how or why it upset her sister. When Kevin didn't clean the litter, it just seemed to clean itself; he didn't appreciate that his neglect made messy work for his parents. And Dad didn't seem to understand how frustrating it is for Mom to be the only caregiver for a newborn.

Game theorists have a name for empathy—they call it *other-regarding preferences*. While the name could use some work, the idea is important. Game theorists (like economists more generally) often get unfairly accused of assuming that people are only selfish money-grubbers. And there is a grain of truth in that accusation—some applications of game theory do make that assumption. But game theory is far more flexible; it can accommodate all sorts of desires and aims. As the game theorist Ken Binmore puts it, "[Game theory] has no difficulty in mod-

eling the kind of saintly folk who would back rather than see a baby cry."

What's exciting about other-regarding they can turn bad interactions into good Take, for example, the Prisoner's Dilem chapter 6—the one where the brother and si way to agree to pick up their LEGOs togethe two siblings to empathize with each other, th to work together becomes self-enforcing. As doesn't care about hurting her brother, she'll part of the deal. But if she cares about her br she'll see how hard he's working and feel bad. N to live up to her promise.

Developing empathy in children is a long and cess, but game theoretic reasoning can help. The game theorist does when she sits down to analyze ation is to see it from all sides. How does each pers the situation? How good or bad will each outcome or her? What considerations go into each outcome?

Psychologists say that by helping your kids see from another's perspective, you will help them to dev pathy. As they start to see how they might react in else's shoes, they start to develop the ability to do this own. This ability develops slowly, starting around four and continuing well into adulthood. And as it develo will start to see how more and more agreements become enforcing.

Dad severely underestimated how often his situation would come up.

So how can you use Kidd's theory to get your kids to care more about the future? Being consistent will help. As children come to expect you to behave in a steady way, they will become confident that those future benefits will really materialize. You might practice delaying rewards at home as well. Ask your child to wait a little bit before getting dessert, and if they do they will get a little more. Or have your kid wait an extra day in order to get a slightly bigger allowance. They might come to expect future rewards to materialize if they see promises come to fruition over and over again.

As your children develop the expectation that future gains will materialize, you will be able to make more self-enforcing agreements. Now your kids will appreciate that it's worth skipping a short-term benefit in order to secure a much larger long-term gain. Once that happens, you can use future rewards to encourage your children to keep their agreements today. Kevin's parents can explain to Kevin that if he keeps his promise to let the cat out every morning, they will be more willing to let him take on responsibilities later. If he's good with the cat, he might be allowed to go out and play by himself next year. Rachel could have explained to her little sister that if she stays out of Rachel's room, then maybe Rachel will let her play with her toys later. And Mom can explain to Dad that if he doesn't keep his promises, in a few months he might not have a wife to make promises to. In all three cases, the reliable prospect of future rewards or punishments makes the agreement self-enforcing.

A lot of ink has been spilled over whether Mischel's or Kidd's way of looking at the Marshmallow Test is correct. As parents, it probably wouldn't hurt to approach your children using both theories. Kidd argues that both explanations are probably involved when your kids decide whether it's worth delaying gratification now for a better payoff in the future. No matter which theory is right, both argue that you can help your kids to forgo short-term excitement for a significant gain in the future.

There is one last trick for making agreements self-enforcing: Teach your children to develop empathy. When children start to care about what happens to other people, beyond themselves, agreements can switch from being unenforceable to self-enforcing. Rachel's sister didn't stay out of Rachel's room, in part, because she failed to appreciate how or why it upset her sister. When Kevin didn't clean the litter, it just seemed to clean itself; he didn't appreciate that his neglect made messy work for his parents. And Dad didn't seem to understand how frustrating it is for Mom to be the only caregiver for a newborn.

Game theorists have a name for empathy—they call it *other-regarding preferences*. While the name could use some work, the idea is important. Game theorists (like economists more generally) often get unfairly accused of assuming that people are only selfish money-grubbers. And there is a grain of truth in that accusation—some applications of game theory do make that assumption. But game theory is far more flexible; it can accommodate all sorts of desires and aims. As the game theorist Ken Binmore puts it, "[Game theory] has no difficulty in modeling the kind of saintly folk who would sell the shirt off their back rather than see a baby cry."

What's exciting about other-regarding preferences is that they can turn bad interactions into good ones very quickly. Take, for example, the Prisoner's Dilemma scenario from chapter 6—the one where the brother and sister couldn't find a way to agree to pick up their LEGOs together. If we can get the two siblings to empathize with each other, then the agreement to work together becomes self-enforcing. As long as the sister doesn't care about hurting her brother, she'll want to shirk her part of the deal. But if she cares about her brother's feelings, she'll see how hard he's working and feel bad. Now, she'll want to live up to her promise.

Developing empathy in children is a long and involved process, but game theoretic reasoning can help. The first thing a game theorist does when she sits down to analyze a social situation is to see it from all sides. How does each person perceive the situation? How good or bad will each outcome be for him or her? What considerations go into each outcome?

Psychologists say that by helping your kids see the world from another's perspective, you will help them to develop empathy. As they start to see how they might react in someone else's shoes, they start to develop the ability to do this on their own. This ability develops slowly, starting around four or five and continuing well into adulthood. And as it develops you will start to see how more and more agreements become self-enforcing.

Our takeaways:

- Help your kids keep their promises by making self-enforcing agreements that the kids will *want* to keep.

- Conventions—promises that are in everyone's short-term interests—are the most powerful self-enforcing agreements. Use them when you can.

- Being consistent with your kids and teaching them to employ "cool thinking" will help them to put off short-term temptations and keep their promises.

- Teaching empathy will help kids see how broken promises hurt other people and help them to develop other-regarding preferences.

9

You Can't Tell Me What to Do!

It's family night and everyone's going out to dinner. In a small miracle, Mom, Dad, Kim, and Max have all agreed to head back home afterward to play board games. That's the end of the good news; now they need to decide where to eat. Dad would prefer Japanese but Italian is a close second. Whatever the family chooses, he desperately wants to avoid yet another round of fast food. Mom wants Italian more than Japanese, but like Dad, she'd like to skip another quick burger. The family decides to vote on where to go. Dad puts in his first pick, and Mom hers. Then, in unison both kids shout "McDonald's!!" It's two to one (to one): McDonald's wins, and Mom and Dad are stuck eating fast food . . . again.

Little did Mom and Dad know the kids saw this one coming. Max likes Burger King more than Kim's favorite, McDonald's. Before the family meeting, Kim pulled Max aside. "Listen, if we disagree about where to go, Mom and Dad will end up choosing.

But, if we present a united front that's two votes for one burger place. They'll have to listen to us." After a short negotiation, Kim and Max agreed: tonight they'll both vote for McDonald's. Next week they'll both vote for Burger King. And Mom and Dad won't know what hit them.

Game theorists have a name for what Kim and Max did. It's called *strategic voting*. Max gamed the system, in a way; he voted for his second favorite option instead of his number one. He didn't get his favorite fries, but at least he got fast food—an outcome that was close to what he wanted. And if he hadn't voted strategically, he might have been stuck eating a spicy tuna roll. Yuck!

Strategic voting is incredibly common in the world of politics—it comes up in almost every primary election in the United States. A left-wing democrat might prefer a socialist to a moderate, but knowing that the socialist could get clobbered in the general election she votes for the moderate—a far better choice than a conservative. Of course, exactly the same calculation (or its mirror image) is being made by those on the right side of the political spectrum. In the end everyone is voting strategically.

Because of complexities like this, game theorists are very interested in understanding the ins and outs of democracy. In fact, some of the earliest discoveries in game theory were made in the eighteenth century, when French mathematicians started thinking about voting. The mathematicians' interest was piqued by the stormy political events of their time: the American and French Revolutions.

Their work launched a long intellectual tradition of careful analysis of voting. It remains a hot topic of discussion today.

And after several centuries of painstaking study, game theorists have come to a disheartening conclusion: Democracy is harder than it looks. In this chapter we'll introduce you to some of the problems game theorists have identified with different forms of democracy, and what the game theorists suggest can be done to solve them. Then you can put these remedies to use in your own little constituency—your family.

Let's start with some good news: Decisions are pretty simple when the family is trying to decide between only two options—whether it's two restaurants, two vacation spots, or two activities. In such situations, game theorists wholeheartedly endorse majority rule: Everyone votes for their favorite option, and the family does whatever the majority chooses.

Game theorists like majority voting for two reasons. When there are only two options, there will never be strategic voting; everyone will be honest when they vote, because no one has an incentive to vote for the option they prefer the least. After all, there's nothing to be gained by it: Voting for your least favorite option can only hurt you. Game theorists have also argued that majority voting is the only "fair" system of deciding. You're now equipped with enough game theory to know that the word "fair" should send up red flags and set off warning bells. So we need to be clear about what we mean. To do that, we must take a short detour to describe how game theorists translate the everyday practice of voting into the language of mathematics.

Here's the idea: Let's assume a kind of electronic ballot box, where everyone goes and casts their vote. Each voter registers whom they would like to vote for by pushing a button on the machine. We all know that there are rules about voting boxes. You can't stuff them with extra ballots (or push the button more than once), even though that's actually happened in American politics. And dead people can't vote, even in Chicago.

Game theorists have their own rules concerning the ballot box. Their ballot box is sophisticated: You vote for your favorite candidate, if you have one, but you can also tell the box that you don't care; either candidate will do. Game theorists begin by assuming they don't know exactly how the ballot box works; they start by describing rules that the ballot box must follow. Then they use mathematics to reach other conclusions about how the box works.

The American mathematician Kenneth May proposed placing four conditions on the metaphorical ballot box. May presented his conditions in the language of mathematics, but here we'll give you the English version:

- Voter equality: The ballot box must treat each person equally. The output of the ballot box shouldn't depend on who is casting a ballot—only on what they voted for. So a system of voting that, say, always gives Mom what she wants is not allowed.
- Neutrality: The ballot box can't be predisposed to any one candidate. So a box that just ignores the votes and spits out "McDonald's" every time won't do. Nor can the ballot box give one option a few extra votes.
- Universal applicability: The vote should always produce a decision. The ballot box can't start smoking and blow up when a particular pattern of votes is cast. It has to work every time. (The voting system *is* allowed to declare a tie, however.)
- Positive association: Here, the winner shouldn't change if more people vote for the winner. Suppose there is a vote on Italian vs. Mexican food for dinner, and Italian wins. If we have another

election where *more* people vote for Italian, then Italian should still win. Said another way, the ballot box shouldn't choose an option because *fewer* people voted for it.

Some of these conditions might strike you as completely obvious; of course that's how a ballot box should work. Who would suggest otherwise? If that's what you thought, good. That was May's point; he thought these are the sorts of conditions that everyone would regard as fair. He then proved that an election decided by a simple majority—where the ballot box spits out the option with the most votes—is the only voting scheme that satisfies these four conditions.

This still might strike you as somewhat obvious—perhaps you already thought that majority voting was obviously the fair way to decide. Or maybe you can see right away how May's conditions equate to simple majority voting. You won't get any argument from us. But don't try to convince any mathematicians; they've been using sledgehammers to kill flies for centuries. It's part of the charm of mathematicians. Take Alfred North Whitehead and Bertrand Russell: They needed 379 pages to prove that one plus one equals two!

Although you might now feel you're smarter than a mathematician, you might have missed an important point: You've probably been violating neutrality all the time at home without knowing it. How? You may have a default option that will "win" if a majority cannot agree on an alternative. The family might stay home and cook a frozen lasagna if they can't agree on a place to eat, or you might go your separate ways if the family can't agree on an activity for the evening. Even the U.S. Congress works this way: If they can't agree on a way to change the law, everything stays the same—no matter how bad that is. This violates neutrality because it gives the status quo a privileged position. Even if everyone agrees that our current situation is terrible, we can't change it unless a majority agrees on a *particular* alternative. (Sound familiar?)

The fact that the existing law is the default has led the American political scientist George Tsebelis to study what he calls *veto players*. Veto players are those people (or political parties) who can prevent a change away from the status quo all by themselves. He argues that the more veto players a country has, the harder it is to change the law. This idea resonates with Paul: When his kids act as veto players, they can never make a decision to go anywhere. (This became more complicated when one of his boys decided McDonald's was no longer his default. Chaos.) While the slow, deliberate way of changing laws might seem appealing to some, it's probably not a productive way to run a family.

The way around this is to be sure that there is never a default option—you are voting on what to do and the outcome of that election stands. No one is a veto player who can unilaterally overturn the result of the decision.

All of this is great when there are only two options for dinner. But what if your town has three "candidate" restaurants? Now the picture gets a little complicated. Once Mom and Dad introduced both Italian and Japanese, they defeated themselves. (They would have done better to engage in a little strategic voting themselves. If they'd both voted Italian, they would have tied the kids and maybe derailed the inevitable run to the Golden Arches.)

Game theorists have been discussing the complexities of voting on multiple options for at least two hundred years. Nicolas de Condorcet was a French philosopher and mathematician and an early champion of the French Revolution. He died somewhat mysteriously after the revolution he had helped start turned against him. Although he was an outspoken champion of democracy, he also invented one of the most troubling problems for our system of government, known today as *Condorcet's paradox.*

To illustrate the problem, let's return to three of our original family members—Mom, Dad, and Kim (Max hasn't been born yet). The family is trying to decide what to do for "family time" after dinner on Friday night. They live in a small town, with only three options: a movie theater, mini-golf, or an activity at home—let's say watching television. Here's how everyone feels about each of the options:

	MOM	DAD	KIM
(FAVORITE)	MOVIE	MINI-GOLF	TELEVISION
(SECOND)	MINI-GOLF	TELEVISION	MOVIE
(LEAST)	TELEVISION	MOVIE	MINI-GOLF

Notice already we have a bit of a problem. Suppose we have a *plurality* election, where everyone votes for their favorite option and the activity with the most votes wins. If we have a three-"candidate," three-voter election and each family member votes for his or her favorite, we have a three-way tie. Mom votes for a movie, Dad votes for mini-golf, and Kim votes for TV. One to one to one.

This might seem easy to solve. Why not break down the complicated decision into a sequence of two-candidate votes?

Mom and Dad will create a "primary." That should be pretty easy for the family. First everyone will vote for either a movie or mini-golf. Then the winner will be put up against TV. College basketball fans will be familiar with brackets like this:

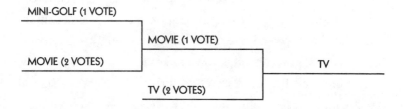

MINI-GOLF (1 VOTE)

MOVIE (1 VOTE)

MOVIE (2 VOTES)

TV

TV (2 VOTES)

When the family votes in the primary, Mom and Kim will vote for a movie over mini-golf, and Dad will be shut out. (Let's suppose for the moment that everyone is voting honestly. No strategic voting.) Mom votes for a movie because it's her favorite activity. Kim votes for it, too, because, although it's not her favorite, she'd prefer it to boring mini-golf. So the movie option wins the primary and advances to the second round to face television. Now Dad and Kim vote for television together, and Mom gets shut out. TV wins far and square, right? Not so fast.

Mom might cry foul. Why did we have to start with a movie and mini-golf in the primary? Why don't we change the system a little? This time, let's have mini-golf and TV face off first. Then the winner will go up against the movie. This produces a different bracket, and paradoxically a completely different outcome.

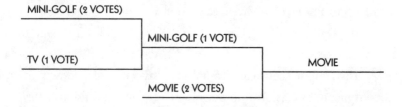

MINI-GOLF (2 VOTES)

MINI-GOLF (1 VOTE)

TV (1 VOTE)

MOVIE

MOVIE (2 VOTES)

Now the family decides to go to a movie! Notice that nobody changed their mind about what they liked; they all feel exactly the same. Nor is anyone voting strategically; they are still being completely honest. But by changing the order we changed the outcome. Now you might already be feeling uncomfortable. It gets worse—if we consider one last way of structuring the election, we get a *third* result:

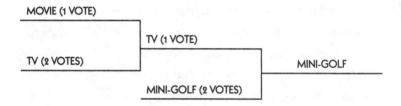

MOVIE (1 VOTE)

TV (1 VOTE)

TV (2 VOTES)

MINI-GOLF

MINI-GOLF (2 VOTES)

This time Dad is ecstatic because they're off for a night of golfing. Now we have a real mess; three different elections, three different outcomes. So instead of arguing about where they're going, the family will sit around and argue over the best way to structure the election. Could they just vote about *that*? That doesn't work because there are three ways to run the election and each of them yields a different outcome. Each member of the family would prefer a different structure for the election. The problem keeps going and going.

Okay, so plurality voting doesn't work to make a decision here; we get a three-way tie no matter what we do. And setting up a primary doesn't work, because the family can't agree on a structure for the primary. Is there anything that does work? It turns out this problem is hard to solve. A pair of mathematical results have shown that when it comes to group decision making, the "best" solution may not be very good at all. Knowing this can help, though, since you'll know not to strive for too much. And we can explain why, somehow, you can never make family decisions. We'll end this chapter by giving you the two best solutions that game theorists have devised for getting around problems you may have already encountered.

The first of these two famous problems for democracy is called *Arrow's impossibility theorem*. It's named after the Nobel Prize–winning economist Kenneth Arrow. Arrow, just like Kenneth May, started off by writing down conditions that an imaginary electronic ballot box should follow. In fact, Arrow came up with that way of thinking about elections first and inspired May. Because it has to deal with elections where there are three (or more) options, Arrow's ballot box is very sophisticated; it takes as input each voter's complete ranking of all the options. So Kim would input that TV was her favorite, followed by a movie, with mini-golf at the bottom. People can even vote for a tie between two options if they like. After recording everyone's vote, the electronic ballot machine then outputs a complete ranking of all the options (with ties, potentially).

Arrow also came up with four conditions that he thought were reasonable operating rules for the ballot-box voting system.

- No dictators: This one is pretty straightforward. It just says that the voting procedure doesn't give one person whatever she wants every time.
- Unanimity: If everybody says mini-golf is better than a movie, then the outcome of the voting machine places mini-golf above the movie.
- Universal applicability: The black box will give us an output no matter what we put into it—no pattern of votes will cause the

machine to break or fail to give us an answer. (This is the same as May's condition above.)

- Independence of irrelevant alternatives: This one is the most complicated, and by far the most controversial. The idea is that if we're just comparing mini-golf to a movie, it doesn't matter how the voters rank some third alternative (like TV). When it's figuring out how to rank mini-golf and a movie, the voting machine sticks only to what the voters say about those two without looking at where the other options fall.

If you have only two options, then majority voting satisfies all four conditions quite easily. But, alas, if you have more than two, you're in trouble. Arrow proved that if you have three options (or more), you cannot build a voting machine that satisfies all four of these conditions. If you satisfy three of them, you must violate the fourth, at least some of the time.

Scholars from fields as diverse as mathematics and philosophy argue over what to do about this. If we want to run an election with three options, we have to violate one of Arrow's conditions—it's a mathematical fact. Which one should we abandon? Some people suggest giving up on universal applicability. Perhaps we don't need a voting machine that works in every situation, only in the most common ones. Voters tend to place themselves on a left–right spectrum—in the United States, at least. Perhaps we could build a black box that only works when voters organize themselves along a simple political spectrum. It turns out you can. But you know what happens to parents who think they can guess what their kids will like. The kids surprise you. And if your system of voting doesn't know how to handle those situations, then you'll be in a whole mess of trouble.

Instead, most scholars suggest giving up the fourth condition—independence of irrelevant alternatives. When you give up *that* condition, many different voting systems become possible. The economist and mathematician Don Saari recommends a voting system named after another eighteenth-century French thinker, Jean-Charles de Borda. You may already use this system without knowing its name—it's called the *Borda count*.

To vote using the Borda count, you have everyone rank all the options—1, 2, 3, and so on. Everyone turns in their scores to Mom and she adds up the rankings for each option. The candidate with the *lowest* total number wins.

To see how this works, let's consider one of those family decisions that always seems to become more stressful with each passing moment of indecision and strife: where to go on vacation. Suppose that Mom, Dad, Kim, and Max have narrowed the possibilities down to three: a theme park, the beach, or a "staycation" in their own city. Mom and Dad both want the staycation—it saves money and will allow them to get some much needed home repairs done in between trips to local museums. But if they have to go somewhere, Mom and Dad want to go to the beach. Kim and Max both hate the idea of the staycation ("Borrrring . . .") But they disagree on favorites: Kim wants to go to the beach, and Max wants Six Flags.

Here's how it all maps out:

	MOM	DAD	MAX	KIM
(FAVORITE)	STAYCATION	STAYCATION	SIX FLAGS	BEACH
(SECOND)	BEACH	BEACH	BEACH	SIX FLAGS
(THIRD)	SIX FLAGS	SIX FLAGS	STAYCATION	STAYCATION

If the family just votes on their favorite options, the staycation wins, with two votes compared to one vote for the theme park and one for the beach. The kids will scream that it's not fair, and for once game theorists will agree with them. The problem here is that by ignoring people's *second* choices, you miss an opportunity for a compromise, and that's what the Borda count picks up on.

If we use the Borda count, here's what we get. Staycation gets two 1's (Mom and Dad) and two 3's (the kids) giving it a total of 8. The theme park doesn't fare as well; it gets two 3's, one 1, and a 2, for a total of 9. (Remember, in the Borda count—like in golf—the lowest number wins.) This means the theme park ranks lower than the staycation. Sorry, Max. We've saved the winner for last. The beach gets three 2's (Mom, Dad, and Max) and one 1 (Kim), giving it a total of 7—and making it a winner. It's only one person's favorite, but everyone can make do.

Ignoring people's second choices can have serious consequences beyond family vacations. Among a whole host of catastrophes, this is one of the problems that plagued the 2000 presidential election in the United States. Many people who voted for Ralph Nader (the Green Party candidate) would rather see the Democrat Al Gore as president than the Republican George W. Bush. Had all the voters who voted for Nader switched to Gore, the Democrats would have won New Hampshire and Florida; this would have turned the election. Because of this result, many people on the political left remain angry at Ralph Nader for ruining the election. Game theorists, on the other hand, might lay the blame on our system of voting.

While many scholars like the Borda count, it does have a flaw: It can be manipulated by a smart voter (or a precocious kid). Let's return to the Condorcet paradox, where Mom, Dad, and Kim are trying to decide what to do on a family outing: mini-golf, a movie, or TV. Now suppose that Kim—a game theorist if there ever was one—suggests they put a fourth option on the ballot: go visit creepy Uncle Larry. Nobody likes Uncle Larry, and it will rank fourth on everyone's list. Here are their *true* preferences with Larry included.

	MOM	DAD	KIM
(FAVORITE)	MOVIE	MINI-GOLF	TELEVISION
(SECOND)	MINI-GOLF	TELEVISION	MOVIE
(THIRD)	TELEVISION	MOVIE	MINI-GOLF
(FOURTH)	UNCLE LARRY	UNCLE LARRY	UNCLE LARRY

If everyone votes honestly, the Borda count isn't terribly helpful here, since it just gives a tie to the three "real" options: the movie, mini-golf, and TV. Each option gets 6 points. Uncle Larry isn't a contender, with 12 points. But, remember, Kim has a plan. Instead of voting honestly, she claims that now she really likes her weirdo uncle. Kim's ballot now reads (1) TV, (2) Uncle Larry, (3) Movie, (4) Mini-golf. Kim didn't change her mind, she's just being sneaky. If Mom and Dad keep voting honestly, but Kim votes strategically, they run into a problem. Now TV wins the Borda count with 6 points. The movie and mini-golf both get 7, and Larry receives a more respectable 10. By introducing a losing option and using it to manipulate the vote, Kim has gotten her favorite option to win instead of being in a tie with two others. With a few more voters or options, we could give you an example in which adding a bad option can promote a sure loser to a winner.

When Jean-Charles de Borda was told about this possibility—that his voting scheme might be manipulated by a wily voter—he exclaimed: "My scheme is intended for honest men!" In the twenty-first century, we should add "honest women." But whatever the sex of the voters, strategic voting spells trouble because you can't always know if your kids (or spouse, for that matter) are being honest. Paul believes *his* children are always honest, of course, but he wouldn't hold others to that exalted standard.

It turns out that the Borda count is in good company: Almost all voting schemes are vulnerable to manipulation by strategic voting. You may recall we mentioned that there is a second bit of mathematical bad news for democracy. This one is called the *Gibbard-Satterthwaite theorem* and is also named for its discoverers: Allan Gibbard, a University of Michigan philosopher, and Mark Satterthwaite, an economist at Northwestern University. Gibbard and Satterthwaite consider all possible *deterministic* designs for our electronic ballot box. "Deterministic" means that the machine never flips a coin at any point—it always gives exactly the same output whenever it gets the same input. What Gibbard and Satterthwaite showed was that every deterministic voting scheme you can come up with is susceptible to strategic voting at least some of the time.

These results are pretty bad news for everyone from presidents to parents. No matter what, we can never design a system for choosing leaders or vacations that is perfect. But it might be a relief to hear this, too: You shouldn't be embarrassed that you can never make family decisions.

While no voting scheme is going to be perfect, game theorists are now trying to figure out if there's a way to say that one voting scheme is better than another. There is a lot of disagreement about how to do this, because there are many different ways to measure "better." (Remember the care we took with the meanings of "fair.") And there are many types of voting.

Instead of jumping into this debate, we're going to point you in another direction. What all of these voting systems have in common is that they want to avoid dictatorships. We had a revolution about this, after all. Beyond dictatorships, these voting schemes also eschew any kind of randomness. People are very averse to the idea that the president of a country might be determined by drawing straws. So scholars like Arrow, Gibbard, and Satterthwaite have focused for the most part on voting schemes that have no dictators and involve no coin flips.

Allan Gibbard does suggest that you might consider going in the opposite direction, however. He suggests two different schemes that explicitly involve flipping a coin to decide some part of the election.

You might be surprised that we'd be suggesting flipping a coin after criticizing it so much in chapter 2. You might try some of the solutions we listed there, too. You could auction off the right to choose the restaurant or vacation. What makes this hard, however, is that it is not always possible to find a fair currency to use. Mom and Dad have a lot more money, and the kids have a lot more free time. So instead, for making family decisions, we'll turn to Gibbard's two proposed solutions. We'll call the first Random Choices and the second Random Dictator.

Random Choices reduces the set of options to just two—where we know that voting works. The family starts with its list of all the options. Dad puts all the options in a hat, and two are drawn at random. These will be the family's choices. Tonight, for example, we'll be deciding between Italian and McDonald's.

Then everyone in the family votes on those two, with a simple majority winning. Gibbard shows that this avoids the pitfall of strategic voting. But it does so with a cost: Sometimes the two options will be unanimously unpopular. What happens when everyone hates Sandy's Sandwiches and Bob's Burgers, but those are the two drawn?

Now, you might be tempted to restrict which of your town's restaurants go in the hat. Perhaps you might let each person write down one restaurant, or you might have a vote to decide which choices go in. But alas, a game theorist wouldn't like that solution. People might strategically choose which restaurants to put in, or vote to remove options from the hat to improve their chances of getting their favorite.

So instead, we recommend you use the Random Dictator scheme. Here's how it works: First, each family member writes down his or her favorite option on a piece of paper. Then all of these ballots are put into a hat, and one is drawn out at random. The option written on that ballot is the winner. This is called Random Dictator because somebody got to be a dictator—whatever they wanted is what happened. But *who gets to be the dictator* is chosen at random.

This would have obvious drawbacks if it were applied to a country the size of the United States. Only a very small fraction of us would ever have the chance to be dictator, and the rest of us would be left completely out of the political process. Gibbard certainly wasn't suggesting that we implement this on a national scale. But for family decisions that are repeated over and over, it wouldn't be so bad.

How would the Random Dictator method work for our dinner decision at the beginning? Remember, Dad wanted Japanese,

Mom wanted Italian, Max liked Burger King, and Kim preferred McDonald's. Kim tricked her parents by getting Max to vote strategically—he voted for McDonald's instead of Burger King. If Mom and Dad had used the Random Dictator scheme, Max would have no reason to vote for McDonald's. If he were chosen to be dictator, he would rather have dictated that the family go to Burger King. If he didn't get chosen, it didn't matter one iota what he voted for. So now, strategic voting doesn't do anyone any good. Mom and Dad still might be stuck with McDonald's, but at least they can take solace in not being outsmarted by their kids.

Our takeaways:

- When there are only two options to choose from, a majority vote is the best way to make a fair decision.
- Beware of default options and veto players. Don't make one choice the default if you can't agree, and don't give anyone a veto.
- Although it has its problems, the Borda count is a good way to make a decision when you have more than two options. But beware of strategic voting.
- Random Dictator is another way to make decisions. It ensures there is never any threat of strategic voting.

Epilogue: Leaving the Nest

Paul and his wife recently invited his older daughter and her fiancé over to dinner. His younger boys dipped in and out of the conversation as the evening progressed. The eight-year-old boy played Minecraft on his Kindle when he wasn't talking to the guests. The five-year-old kept dragging them into his room to show them models he'd made at summer camp and games he wanted them to play. When Paul asked him gently to let the guests sit and eat their dinner, his son demurred. "When somebody comes over, you have to be nice," his son said. The boy wasn't trying to distract them from their dinner. He was engaging them and showing them some old-fashioned hospitality.

Paul and his wife were delighted. Their son, without any direction from them, had decided on his own that sharing his crafts and toys was a nice way to treat dinner guests. Paul and his wife couldn't remember making a point of this. Had they done something right without realizing it? Or did they just get lucky on this one?

Game theory has a lot that's useful to say about one of the most important transitions that children will make in their lives—the move from the family to the wider world. Paul's younger kids have some time before they make that move, but it's not too soon to start thinking about what tools our kids will need to function in that wider world. Let's examine this bridge from the family to society and discuss what we can and should do to help prepare our children to cross that span. For many parents, this is close to the last stop. It's time to start letting go.

Before we think about preparing our children to adapt to the norms of their culture and society, we should be clear on what social norms are. Some can be undesirable (as we will see with the practice of foot-binding in China). But many encourage good behavior, and those are the norms for which we'd like to prepare the kids. We find these prosocial norms even in the realm of economics—and we have no less an authority for that than the great political economist and moral philosopher Adam Smith. He gave us one of the simplest and clearest descriptions of a free market in his book *The Wealth of Nations*, published in 1776:

"It is not from the benevolence of the butcher, the brewer, or the baker, that we expect our dinner, but from their regard to their own interest," he wrote. They sell you fresh, tasty food not because they want you to enjoy your meal, but because they want to keep you coming back. That's the way they prosper. If all of our dealings with others were this simple and straightforward, we could merely instruct our children to follow their own interests and all would be well. But the pursuit of self-interest doesn't explain all of human behavior, as many of our examples have shown.

Smith himself made that clear in an earlier book, *The Theory of Moral Sentiments*: "How selfish soever man may be supposed, there are evidently some principles in his nature, which interest him in the fortune of others, and render their happiness necessary to him, though he derives nothing from it, except the pleasure of seeing it." Aha! There we see a quality we would like to encourage, and the way we do that is by establishing norms of behavior that reflect sensitivity and empathy.

As an indication of how powerful social norms can be, let's look at a notorious example of a social norm gone dangerously wrong—the former practice in China of binding girls' feet to keep them small. Some of the earliest evidence for foot-binding comes from the tomb of Lady Huang Sheng, the wife of an imperial clansman, who died in 1243, the historian Amanda Foreman has written. The practice "is said to have been inspired by a tenth-century court dancer named Yao Niang who bound her feet into the shape of a new moon. She entranced Emperor Li Yu by dancing on her toes inside a six-foot golden lotus festooned with ribbons and precious stones. In addition to altering the shape of the foot, the practice also produced a particular sort of gait that relied on the thigh and buttock muscles for support. From the start, foot-binding was imbued with erotic overtones. Gradually, other court ladies—with money, time and a void to fill—took up foot-binding, making it a status symbol among the elite," Foreman writes. The practice persisted for a thousand years, finally disappearing in the twentieth century as tradition gradually gave way to progress. This odd historical episode shows how difficult it can be to change social norms, and how long they can persist, even if the social norm requires an act as barbaric as foot-binding.

Cristina Bicchieri, a philosopher and game theorist at the University of Pennsylvania, has spent a long time studying the origins and nature of social norms. She has called social norms "the grammar of society," and she makes it clear that norms come in multiple forms. Social norms, she writes, "are the language a society speaks, the embodiment of its values and collective desires, the secure guide in the uncertain lands we all traverse, the common practices that hold human groups together." Some norms begin accidentally. Teenagers chance to wear some of the same clothes, and soon the clothes become imbued with meaning, and more and more teenagers want to adopt the norm—they want to wear what everyone else is wearing. And some norms spring up to challenge existing norms. In the 1960s, short hair was the norm for men. The Beatles created a cultural revolution with their "long" haircuts that are so tame by today's standards that it's hard to understand why they caused so much comment—until you realize that it was because the Beatle cuts violated a social norm.

As parents, we might be concerned about what our teenagers are wearing, but their clothes are not always tied to prosocial or antisocial behavior. Sometimes a hat is just a hat. We should be more concerned with significant norms that relate to prosocial behavior. These behaviors, Bicchieri writes, include "norms of cooperation, promise-keeping, reciprocity, or fairness," which "are some of the institutions that allow a society to function smoothly." They might even be necessary for a society's existence, she writes. Indeed, prosocial behavior—offering a fair split to someone when playing the Ultimatum Game, for example—existed before we had institutions to punish departure from social norms. Fairness arose spontaneously. (The Ultimatum Game, remember, is the one in which a person has to divide money with another, and must impress the recipient as fair or risk both of them losing everything.) "The propensity to recognize and conform to norms, as well as to be prepared to punish defectors, is evolutionarily necessary to the production and maintenance of any public good," Bicchieri writes. And yet, attempts to cheat—what we might call "defection without detection"—are also older than humanity.

Bicchieri has also looked at the way children respond to social norms. She and her colleagues have found that children are not quite as sophisticated as adults, but that they are already beginning to use their knowledge of social norms strategically. But adults' behavior varies in connection with whether they think they will be caught by others for violating the norms. "Individuals tend to comply with the norm when risking sanctions, but disregard the norm when violations are undetectable," Bicchieri and her colleagues write. Children (mercifully for parents!) are less advanced in their reasoning about when their violations of norms can or cannot be detected.

Bicchieri and her colleagues asked children eight to ten years old to play the Ultimatum Game. The researchers found that children made decisions partly on the basis of how much they thought others knew about their choices. And they would try to make use of what others knew in a self-serving way. That is, kids would choose to be slightly unfair when they could blame the unfairness on some outside force—in this case the flip of a coin.

The UCLA anthropologist and game theorist Joan Silk has likewise run game-theory tests and other experiments with kids to try to understand the psychology of prosocial behavior. Her research suggests that an awareness of social norms begins at an early age. "People are generous and seem to care about their reputations," Silk told us. "They care about these things even when they don't know the other people . . . People hold doors open, and go to potluck dinners without taking all the food for themselves." We shouldn't be surprised by this, because we've already seen how empathy and concern play important roles in our families. But the leap to cooperating with others is part of what makes us unique. Silk has looked for evidence of empathy and cooperation in other primates, and she doesn't find nearly as much. Nonhuman primates, including the great apes, mostly limit their cooperation to kin, mates, and partners who reciprocate. We are, Silk has written, "unusual apes." We extend cooperation to strangers, and we do it often.

The question Silk asked was, How did we get that way?

"We talk in kindergarten about sharing, and the idea is, kids would never do this unless we taught them." Silk wanted to study that. "Is that true, or do they have a naturally developing tendency to be generous? It may be that the messages we're sending to three- and five-year-olds aren't having much impact. But they might have more impact when they are older."

She and her colleagues ran a series of games with 326 children three to fourteen years old from six societies, and 120 adults from five of those societies. The participants included hunter-gatherers from the Congo basin, farmers from Namibia and the Amazon, foragers from Australia and Melanesia, and residents of Los Angeles. Among the games studied was the

Costly Sharing Game, in which participants were given the choice of keeping two servings of food or sharing one with another. Sharing had a cost. In the Prosocial Game, participants could choose between one portion for themselves or one for themselves plus one for somebody else. This was sharing without any cost for the participant. When Silk and her colleagues had run a series of these games and others, and analyzed the data, they made a surprising discovery: Costly prosocial behavior was more common in young children than in older kids.

In three- and four-year-olds from different cultures, there was little difference in prosocial behavior. Apparently the kids were not greatly affected by their culture's social norms. But by middle childhood, around ages seven and eight, they had begun reflecting the values of their society. Kids from Los Angeles now began to look different from kids in the Congo or the Amazon. Each reflected the norms of their own culture.

The ages at which these changes occur should sound familiar; we've seen them before. Early childhood is the time at which children are only beginning to be sensitive to inequity. Middle childhood is when they begin to be uncomfortable with inequity even in their favor. "Middle childhood may be when children begin to conform to cooperative social norms, even if they may have learned these norms years prior," Silk and her colleagues write. The biggest cultural differences were seen in the Costly Sharing Game, in which children share some of their own candy and so pay a cost to do what's expected of them—that is, to share. The experiments showed that generosity without a personal cost increased steadily with age in children of all cultures. But older children tracked their particular cultures' adult social norms regarding sharing that was costly. The development of prosocial behavior is the product of "a complex interaction with acquired local culture," the researchers write.

If we've done our work, using game theory and all of our wits to encourage our children to behave in the ways we'd like, this transformation should happen as a matter of course. We shouldn't have to resort to scenarios like this one from Jonathan Burke, who teaches game theory at Pepperdine University: "I tell students, after they have their first few kids, to Photoshop into their family pictures a slightly older child. When their kids ask about that child, say, 'Ah, poor Phillip, we loved him. But one day he tracked mud into the room and then lied about it! We really miss him.'" Paul and his wife have been tempted to try this, but the kids usually redeem themselves just in time.

Silk has an optimistic view of kids' adoption of social norms. "This is a lesson kids can learn, and if we value generosity, both modeling it and teaching it is effective . . . We think our kids are more selfish than they are. And they aren't. Our kids are nice. We're doing a good job. They're very responsive to the messages we're sending—you know, be nice. They're not just selfish machines; they're built to care for other people . . . If we don't screw it up, our kids will grow up to be perfectly nice people.

"The trick is to explain to them why they will have a better life if they act as moral people, because the short-term costs of being a moral person, of not being selfish, have long-term benefits."

As we said at the outset, game theory's perspectives on children and families are not meant to turn us into schemers or set us up to joust with or otherwise game our children. Game theory helps us understand why it can make sense to share, to be honest, and to cooperate. We can play the Ultimatum Game, the Prisoner's Dilemma, and other game-theory games for years with our children before releasing them into the wild. Game theory can help us know why our kids behave the way they do. And it shows us what we can do to ease tensions in our family and to teach our kids some of the notions of fairness, generosity, and sharing that make them "nice," as Silk says.

We encourage you to play these games with your children. We hope we will help you to ease some of the conflict in the family, or maybe even eliminate it. And we hope we will help you to prepare your children for the world outside the family. You might find that an understanding of game theory will help you, too—at work, and in your dealings with extended-family members and friends.

Make a practice of watching your children play their games, and enjoy the games you play with them. Delight in the games you and your partner play as you try to grapple with the charms and challenges of childrearing. And most of all, enjoy your children. They are good little strategists. And they are nice people.

Notes

Introduction

4 *After the contest, Sotheby's acknowledged*: Carol Vogel, "Rock, Paper, Payoff: Child's Play Wins Auction House an Art Sale," *The New York Times*, April 29, 2005, www.nytimes.com/2005/04/29/arts/design/29scis .html?_r=0.

4 *One famous example*: https://en.wikipedia.org/wiki/Zhuge_Liang.

7 *Spiders and fish, it turns out*: Ken Binmore, *Game Theory: A Very Short Introduction* (New York: Oxford University Press, 2007), 2.

1. I Cut, You Pick

14 *the poem* Theogony, *by Hesiod*: The *Theogony* of Hesiod, trans. Hugh G. Evelyn-White (1914), www.sacred-texts.com/cla/hesiod/theogony.htm.

14 *Prometheus should have been smart*: Ibid.

15 *Lot chose the plain of Jordan*: Genesis 13:10–11, www.kingjamesbible online.org/Genesis-Chapter-13/.

15 *the writer Primo Levi recalled*: Steven J. Brams and Alan D. Taylor, *The Win-Win Solution: Guaranteeing Fair Shares to Everybody* (New York: W.W. Norton and Company, 1999), 53–54.

16 *produce an envy-free division*: Hugo Steinhaus, "The Problem of Fair Division," *Econometrica* 16, no. 1 (1948): 101–104.

17 *the British would have incentive*: Brams and Taylor, 44.

20 *the father of game theory*: Len Fisher, *Rock, Paper, Scissors: Game Theory in Everyday Life* (New York: Basic Books, 2008), 36.

22 *"madmen like Hitler or Stalin"*: Binmore, 2.

23 *"seen to be so greedy"*: Fisher, 38.

25 *both would be worse off*: Brams and Taylor, 61.

25 *it took forty years or so*: Fisher, 48.

28 requires *really complicated math*: Jack Robertson and William Webb, *Cake-Cutting Algorithms: Be Fair If You Can* (Natick, MA: A. K. Peters, 1998).

28 *a website to do the division*: www.spliddit.org.

2. Don't Cut Barbie in Half!

35 *In economics there is a distinction*: Alon Harel, Zvi Safra, and Uzi Segal, "Ex-Post Egalitarianism and Legal Justice," *Journal of Law, Economics, and Organization* 21 (2005): 57–75, doi:10.1093/jleo/ewi003.

37 *Didius Julianus won with his bid*: Cassius Dio, *Roman History*, trans. Earnest Cary (Cambridge: Harvard University Press, 1914–1927), Book LXXIV, Chapter 11, accessed August 13, 2015, http://penelope.uchicago.edu/Thayer/E/Roman/Texts/Cassius_Dio/74*.html.

37 *executed sixty-six days later*: Cassius Dio, chapter 17.

37 *The Stockholm Auction House*: "Stockholms Auktionsverk—the World's Oldest Auction House," Stockholms Auktionsverk, accessed August 13, 2015, http://auktionsverket.com/about-us/about-stockholms-auktionsverk/.

37 *Christie's, the world's largest auction house*: Mark Odell and Duncan Robinson, "Christie's Posts Record Profits as First-Half Revenues Hit £2.7bn," *Financial Times*, July 16, 2014, www.ft.com/intl/cms/s/0/f3c4d74e-0ccf-11e4-bf1e-00144feabdc0.html#axzz3iicplOGl.

39 *when he wrote a paper on the subject*: William Vickrey, "Counterspeculation, Auctions, and Competitive Sealed Tenders," *Journal of Finance* 16, no. 1 (1961): 8–37.

39 *everyone submit bids by mail*: David Lucking-Reiley, "Vickrey Auctions

in Practice: From Nineteenth-Century Philately to Twenty-First-Century E-Commerce," *Journal of Economic Perspectives* 14, no. 3 (2000): 183–92, doi:10.1257/jep.14.3.183.

41 *economists are busy debating*: John H. Kagel and Dan Levin, "Auctions (Experiments)," from *The New Palgrave Dictionary of Economics*, 2nd ed., eds. Steven N. Durlauf and Lawrence E. Blume (Palgrave Macmillan, 2008), *The New Palgrave Dictionary of Economics Online*, accessed August 13, 2015, www.dictionaryofeconomics.com/article?id=pde2008 _A000241&goto=auctions&result_number=157.

41 *the* revenue equivalence theorem: Paul Klemperer, *Auctions: Theory and Practice*, The Toulouse Lectures in Economics (Princeton: Princeton University Press, 2004).

42 *This auction became popular in England*: William S. Walsh, *A Handy Book of Curious Information: Comprising Strange Happenings in the Life of Men and Animals, Odd Statistics, Extraordinary Phenomena and Out of the Way Facts Concerning the Wonderlands of the Earth* (Philadelphia: J. B. Lippincott Co., 1913): 63–64.

43 *Dutch auctions were introduced*: Mike Dash, *Tulipomania: The Story of the World's Most Coveted Flower and the Extraordinary Passions It Aroused* (New York: Crown, 2000).

43 *still used in the Netherlands*: "Location: Flower Auction—Aalsmeer," Regal Travel Service, accessed August 13, 2015, http://regaltravel.nl /places-of-interest/flower-auction---aalsmeer.

43 *used by the U.S. Treasury*: "Treasury Announces Intent to Sell Preferred Stock in Public Dutch Auction," U.S. Department of the Treasury Press Center, September 5, 2012, www.treasury.gov/press-center /press-releases/Pages/tg1697.aspx.

43 *a way to price companies*: Telis Demos, "Exactly What Is a Dutch Auction?" *Deal Journal, Wall Street Journal* Blogs, July 21, 2012, http:// blogs.wsj.com/deals/2012/06/21/exactly-what-is-a-dutch-auction/.

45 *The computer scientist Ruggiero Cavallo*: Ruggiero Cavallo, "Optimal Decision-Making with Minimal Waste: Strategyproof Redistribution of VCG Payments," *AAMAS '06: Proceedings of the Fifth International Joint Conference on Autonomous Agents and Multiagent Systems* (New York: ACM, 2006), 882–89, http://doi.acm.org/10.1145/1160633.1160790.

45 *based on a complex formula*: Efthymios Athanasiou, "A Solomonic

Solution to the Problem of Assigning a Private Indivisible Good," *Games and Economic Behavior* 82 (2013): 369–87, doi:10.1016/j.geb .2013.07.007.

3: He Got a LEGO Set? That's Not Fair!

48 *"everyone else was hit"*: J. R. Benjamin, "Sidney Morgenbesser's Sense of Humor," *The Bully Pulpit*, October 10, 2013, http://jrbenjamin.com /2013/10/10/sidney-morgenbessers-sense-of-humor/.

52 *Even when the money gets big*: Lisa A. Cameron, "Raising the Stakes in the Ultimatum Game: Experimental Evidence from Indonesia," *Economic Inquiry* 37, no. 1 (1999): 47.

53 *isn't even close to the beginning*: Frans de Waal, *The Age of Empathy: Nature's Lessons for a Kinder Society* (New York: Crown, 2009), 185.

53 *13 million years ago*: Dan Vergano, "Ancient Human-Chimp Link Pushed Back Millions of Years," *National Geographic* News, June 12, 2014, http://news.nationalgeographic.com/news/2014/06/140612-chimp -father-evolution-human-science/.

54 *"those paltry cucumber slices"*: de Waal, 187.

58 *they can't explain why*: Jonathan Haidt, "The Emotional Dog and Its Rational Tail: A Social Intuitionist Approach To Moral Judgment," *Psychological Review* 108, No. 4 (2001): 814.

59 *others have minds of their own*: Alison Gopnik, "How Babies Think," *Scientific American*, July 2010, 76–81, www.alisongopnik.com/papers _alison/sciam-gopnik.pdf.

61 *"succeeded rather than failed"*: Mark Sheskin et al., "Anti-Equality: Social Comparison in Young Children," *Cognition* 130, no. 2 (2014): 153, www.ncbi.nlm.nih.gov/pmc/articles/PMC3880565/..

61 *his colleagues drily observe*: Sheskin et al.

62 *scandalously unscientific title*: Peter R. Blake and Katherine McAuliffe, "I Had So Much It Didn't Seem Fair," *Cognition* 120, no. 2 (2011): 215–24.

63 *they had more aces*: Judith Mehta, Chris Starmer, and Robert Sugden, "An Experimental Investigation of Focal Points in Coordination and Bargaining: Some Preliminary Results," in *Decision Making Under Risk and Uncertainty: New Models and Empirical Findings*, ed. John Geweke

(Dordrecht, Netherlands: Springer Science+Business Media, 1992), 211–19.

64 *clearly benefit all parties*: Linda Babcock and George Loewenstein, "Explaining Bargaining Impasse: The Role of Self-Serving Biases," *Journal of Economic Perspectives* 11, no. 1 (1997): 109–26, doi:10.1257/ jep.11.1.109.

68 *mainly to keep the peace*: de Waal, 187.

4: *You Can't Be Serious!*

72 *The idea of credible*: Reinhard Selten, "Spieltheoretische Behandlung eines Oligopolmodells mit Nachfrageträgheit: Teil I: Bestimmung des Dynamischen Preisgleichgewichts," *Zeitschrift für die Gesamte Staatswissenschaft* 121 (1965), 301–24.

72 *the Chain Store Game*: Reinhard Selten, "The Chain Store Paradox," *Theory and Decision* 9, no. 2 (1978): 127–59.

75 *Interesting paradoxes and problems*: Philip J. Reny, "Rationality in Extensive-Form Games," *The Journal of Economic Perspectives* 6, no. 4 (1992): 103–18.

77 *intended to counter any aggression*: Steven Metz, *Eisenhower as Strategist: The Coherent Use of Military Power in War and Peace* (Carlisle, PA: Strategic Studies Institute, 1993), www.strategicstudiesinstitute .army.mil/pubs/summary.cfm?q=359.

78 *Kahn floated the idea*: Herman Kahn, *On Thermonuclear War* (Princeton, NJ: Princeton University Press, 1960).

78 *Kubrick met with Kahn*: Sharon Ghamari-Tabrizi, *The Worlds of Herman Kahn: The Intuitive Science of Thermonuclear War* (Cambridge, MA: Harvard University Press, 2005).

79 *Yu's strategy worked*: "Xiang Yu," *Wikipedia*, accessed August 13, 2015, https://en.wikipedia.org/wiki/Xiang_Yu.

79 *Tariq ibn Ziyad*: "Tariq ibn Ziyad," *Wikipedia*, accessed August 13, 2015, https://en.wikipedia.org/wiki/Tariq_ibn_Ziyad.

82 *Hobbes argued that people*: Thomas Hobbes, *Leviathan: With Selected Variants from the Latin Edition of 1668*, ed. Edwin Curley, Hackett Classics (Indianapolis, IN: Hackett Publishing Company, 1994 [first published 1651]).

82 *threats can become credible*: George J. Mailath and Larry Samuelson, *Repeated Games and Reputations: Long-Run Relationships* (New York: Oxford University Press, 2006), chapter 17.

85 *"rewards" and "punishments" differently*: Amos Tversky and Daniel Kahneman, "The Framing of Decisions and the Psychology of Choice," *Science* 211 (1981): 435–58.

85 *"no one will be saved"*: Tversky and Kahneman, 435.

86 *"people will die"*: Ibid.

86 *The game theorist Joshua Weller*: Joshua A. Weller, Irwin P. Levin, and Natalie L. Denburg, "Trajectory of Risky Decision Making for Potential Gains and Losses from Ages 5 to 85," *Journal of Behavioral Decision Making* 24, no. 4 (2011): 331–44, doi:10.1002/bdm.690.

5. The Dog Ate My Homework

90 *from bacteria to orangutans*: John Maynard Smith and George R. Price, "The Logic of Animal Conflict," *Nature* 246 (1973): 15–18.

91 *the bigger female eats Photinus*: Thomas Eisner, *For Love of Insects* (Cambridge, MA: The Belknap Press, 2003), 141ff.

92 *conflicts with Mom and Dad*: Robert L. Trivers, "Parent-Offspring Conflict," *American Zoologist* 14, no. 1 (1974): 249–64.

93 *this harassment sometimes works*: Stephen T. Emlen and Peter H. Wrege, "Parent-Offspring Conflict and the Recruitment of Helpers among Bee-eaters," *Nature* 356 (1992): 331–33.

93 *their provocation rarely succeeds*: Caroline E. G. Tutin, "Responses of Chimpanzees to Copulation, with Special Reference to Interference by Immature Individuals," *Animal Behaviour* 27, Part 3 (1979): 845–54.

94 *hungrier than it really is*: William A. Searcy and Stephen Nowicki, *The Evolution of Animal Communication: Reliability and Deception in Signaling Systems* (Princeton, NJ: Princeton University Press, 2005), 50–51.

95 *what biologists call an* index: John Maynard Smith and David Harper, *Animal Signals* (New York: Oxford University Press, 2003).

96 *heavier spiders create greater perturbations*: Susan E. Riechert, "Games Spiders Play III: Cues Underlying Context-Associated Changes in Agonistic Behavior," *Animal Behaviour* 32 (1984): 1–15.

97 *were punished for lying*: Anders Pape Møller, "Social Control of Deception

Among Status Signalling House Sparrows *Passer domesticus*," *Behavioral Ecology and Sociobiology* 20, no. 5 (1987): 307–11, doi:10.1007/BF00300675; Searcy and Nowicki.

99 *"even when he speaks the truth"*: Joseph Jacob, adapter, *The Fables of Æsop: Selected, Told Anew, and Their History Traced* (London: MacMillan and Co., 1922 [first published 1894]), 102–103.

100 *in a situation of perfect monitoring*: Mailath and Samuelson.

100 *about how hungry they are*: R. E. Ricklefs, "The Roles of Parent and Chick in Determining Feeding Rates in Leach's Storm-Petrel," *Animal Behaviour* 43, no. 6 (1992): 895–906, doi:10.1016/S0003-3472(06)80003-5.

102 *as her loyal servant*: Mike Dixon-Kennedy, *Encyclopedia of Greco-Roman Mythology* (Santa Barbara, CA: ABC-CLIO, 1998).

103 *in the wild, the handicap principle*: Amotz Zahavi and Avishag Zahavi, *The Handicap Principle: A Missing Piece of Darwin's Puzzle* (New York: Oxford University Press, 1997).

104 *given it mathematical precision*: Alan Grafen, "Biological Signals as Handicaps," *Journal of Theoretical Biology* 144 (1990): 517–46.

104 *Some game theorists (including Kevin)*: Kevin J. S. Zollman, "Finding Alternatives to Handicap Theory," *Biological Theory* 8 (2013), 127–32, doi:10.1007/s13752-013-0107-1.

105 *simply recalling the past*: Aldert Vrij, Pär Anders Granhag, Samantha Mann, and Sharon Leal, "Outsmarting the Liars: Toward a Cognitive Lie Detection Approach," *Current Directions in Psychological Science* 20, no. 1 (2011): 28–32, doi:10.1177/0963721410391245.

107 *haven't evolved ways of dealing with it*: Robert B. Payne, *The Cuckoos*, Bird Families of the World, Book 15 (New York: Oxford University Press, 2005).

6: He Started It!

111 *"the group life of these monkeys"*: de Waal, 171.

111 *an estimated 2.5 billion insects*: "Facts About This Colony," Florida Museum of Natural History, www.flmnh.ufl.edu/bats/facts-about-colony/.

112 *what you might call friends*: Robert Wright, *The Moral Animal: Why We Are the Way We Are: The New Science of Evolutionary Psychology* (New York: Pantheon 1994), 203.

112 indirect reciprocity: Martin A. Nowak and Karl Sigmund, "Evolution

of Indirect Reciprocity by Image Scoring," *Nature* 393 (1998): 573–77, www.nature.com/nature/journal/v393/n6685/full/393573a0.html.

113 *the environment in their families*: Kristina R. Olson and Elizabeth S. Spelke, "Foundations of Cooperation in Young Children," *Cognition* 108 (2008): 222–31, https://depts.washington.edu/uwkids/olson.spelke .2008.pdf.

117 *"an ongoing interaction with another"*: Robert Axelrod, *The Evolution of Cooperation*, revised edition (New York: Basic Books, 1984), vii.

117 *"egoists without central authority"*: Axelrod, 3.

119 *Once again, Tit for Tat won*: Axelrod, viii.

120 *encourage the other one*: Jennifer Breheny Wallace, "Game Theory Secrets for Parents," *The Wall Street Journal*, July 10, 2014, www.wsj.com /articles/game-theory-secrets-for-parents-1405005848.

121 *established, it often stuck*: Axelrod, 21.

121 *"logic of kill or be killed"*: Axelrod, 74.

123 *"to heal all breaches"*: P. G. Wodehouse, *Just Enough Jeeves: Right Ho, Jeeves; Joy in the Morning; Very Good, Jeeves* (New York, W. W. Norton, 2010), 710, as quoted in Fisher, 181–82.

124 *"restore mutual cooperation almost immediately"*: Jianzhong Wu and Robert Axelrod, "How to Cope with Noise in the Iterated Prisoner's Dilemma," *The Journal of Conflict Resolution* 39, no. 1(1995): 183–89.

124 *Spite—the "shady relative"*: Rory Smead and Patrick Forber, "The Evolutionary Dynamics of Spite in Finite Populations," *Evolution: International Journal of Organic Evolution* 67, no. 3 (2013): 698–707, doi:10.1111/j.1558-5646.2012.01831.x.

125 *paper published in 2014*: Katherine McAuliffe, Peter R. Blake, and Felix Warneken, "Children Reject Inequity out of Spite," *Biology Letters* 10, no. 12 (S014): 20140743, http://dx.doi.org/10.1098/rsbl.2014.0743.

7: Why Can't *You* Pay for This?

133 *produce as much as possible*: Jean-Jacques Laffont and David Martimort, *The Theory of Incentives: The Principal-Agent Model* (Princeton, NJ: Princeton University Press, 2002).

134 *more than she is willing*: Binmore, 105–106.

134 *a technical term for this*: Avinash Dixit, Susan Skeath, and David H. Reiley,

Jr., *Games of Strategy*, 4th ed. (New York: W. W. Norton and Company, 2014), 281.

138 *"takes longer to mature"*: Sarah Blaffer Hrdy, *Mothers and Others: The Evolutionary Origins of Mutual Understanding* (Cambridge, MA: The Belknap Press, 2009), 146.

142 *the rule will evaporate*: Finn E. Kydland and Edward C. Prescott, "Rules Rather than Discretion: The Inconsistency of Optimal Plans," *The Journal of Political Economy* 85, No. 3 (1977): 473–92.

8. Are You Saying You Don't Believe Me?

147 *more than four thousand years*: Lloyd Duhaime, "2550 BC—The Treaty of Mesilim," Duhaime's Timetable of World Legal History, last updated May 5, 2012, www.duhaime.org/LawMuseum/LawArticle-1313/2550-BC--The-Treaty-of-Mesilim.aspx.

147 *well over a century*: "Origins of Organized Crime," Crime Museum, accessed August 13, 2015, www.crimemuseum.org/crime-library/origins-of-organized-crime.

148 *form groups in the first place*: Peter J. Richerson and Robert Boyd, *Not by Genes Alone: How Culture Transformed Human Evolution* (Chicago: University of Chicago Press, 2005).

149 *one hundred years before Nash*: Antoine Augustin Cournot, *Recherches sur les Principes Mathématiques de la Théorie des Richesses* (Paris: Chez L. Hachette, 1838).

150 *playing a zero-sum game*: John von Neumann and Oskar Morgenstern, *Theory of Games and Economic Behavior* (Princeton, NJ: Princeton University Press, 1944; sixtieth-anniversary ed., 2004).

150 *Nash showed that the equilibrium*: John F. Nash, "Equilibrium Points in N-Person Games," *Proceedings of the National Academy of Science* 36, no. 1 (1950): 48–49.

151 *If you're a fan of the movie*: Bart Kosko, "How Many Blonds Mess Up a Nash Equilibrium?" *Los Angeles Times*, February 13, 2002, http://articles.latimes.com/2002/feb/13/opinion/oe-kosko13.

154 *Hume gave us many examples*: David Hume, *A Treatise of Human Nature: Being an Attempt to Introduce the Experimental Method of Reasoning into Moral Subjects* (London: John Noon, 1739), Book 3.

155 *Adam Smith posited the "invisible hand"*: Adam Smith, *An Inquiry into the Nature and Causes of the Wealth of Nations* (London: W. Strahan and T. Caddell, 1776).

155 *Hume and Smith were friends*: "The Death of David Hume: Letter from Adam Smith, LL.D. to William Strachan, Esq.," accessed August 13, 2015. www.ourcivilisation.com/smartboard/shop/smitha/humedead .htm.

155 *modern philosophers, such as Brian Skyrms*: Brian Skyrms, *Evolution of the Social Contract* (Cambridge, UK: Cambridge University Press, 1996).

157 *There have been countless experiments*: Shane Frederick, George Loewenstein, and Ted O'Donoghue, "Time Discounting and Time Preference: A Critical Review," *Journal of Economic Literature* 40, no. 2 (2002): 351–401, doi:10.1257/002205102320161311.

158 *in a book published in 1834*: John Rae, *Statement of Some New Principles on the Subject of Political Economy, Exposing the Fallacies of the System of Free Trade, and of Some Other Doctrines Maintained in the "Wealth of Nations"* (Boston: Hilliard, Gray, & Co., 1834).

158 *publicized his work many years later*: R. Warren James, "Rae, John (1796–1872)," in *Dictionary of Canadian Biography*, vol. 10 (University of Toronto/Université Laval, 2003), accessed August 13, 2015, www .biographi.ca/en/bio/rae_john_1796_1872_10E.html.

159 *The well-known Marshmallow Test*: Walter Mischel and Ebbe B. Ebbesen, "Attention in Delay of Gratification," *Journal of Personality and Social Psychology* 16, no. 2 (1970): 329–37, doi:10.1037/h0029815.

160 *in a number of other areas*: Walter Mischel, Yuichi Shoda, and Monica L. Rodriguez, "Delay of Gratification in Children," *Science* 244 (1989), 933–38, doi:10.1126/science.2658056.

161 *uncertain amount of yumminess*: Leonard Green, Joel Myerson, and Pawel Ostaszewski, "Discounting of Delayed Rewards across the Life Span: Age Differences in Individual Discounting Functions," *Behavioural Processes* 46 (1999), 89–96, doi:10.1016/S0376-6357(99)00021-2.

162 *a variation on the marshmallow experiment*: Celeste Kidd, Holly Palmeri, and Richard N. Aslin, "Rational Snacking: Young Children's Decision-Making on the Marshmallow Task is Moderated by Beliefs about Environmental Reliability," *Cognition* 126 (2013), 109–14, doi:10 .1016/j.cognition.2012.08.004.

165 *"rather than see a baby cry"*: Ken Binmore, *Rational Decisions* (Princeton, NJ: Princeton University Press, 2009), 8.

9. You Can't Tell Me What to Do!

171 *satisfies these four conditions*: Kenneth O. May, "A Set of Independent Necessary and Sufficient Conditions for Simple Majority Decision," *Econometrica* 20, no. 4 (1952), 680–84, doi:10.2307/1907651.

171 *one plus one equals two*: Alfred North Whitehead and Bertrand Russell, *Principia Mathematica* Volume I, first edition (Cambridge: Cambridge University Press, 1910), 379.

171 *a way to change the law*: George Tsebelis, *Veto Players: How Political Institutions Work* (Princeton, NJ: Princeton University Press, 2002).

173 *the revolution he had helped start*: David Williams, *Condorcet and Modernity* (New York: Cambridge University Press, 2004), 42–43.

173 Condorcet's paradox: James Stodder, "Strategic Voting and Coalitions: Condorcet's Paradox and Ben-Gurion's Tri-lemma," *International Review of Economics Education* 4, no. 2 (2005): 58–72, doi:10.1016/S1477 -3880(15)30131-6.

176 Arrow's impossibility theorem: Kenneth J. Arrow, *Social Choice and Individual Values* (New York: John Wiley and Sons, 1951; repr. New Haven, CT: Yale University Press, 1970).

177 *It turns out you can*: Duncan Black, *The Theory of Committees and Elections* (New York: Cambridge University Press, 1958; repr. Kluwer Academic Publishers, 1987).

178 *Saari recommends a voting system*: Donald G. Saari, "The Symmetry and Complexity of Elections," accessed August 13, 2015, www.colorado .edu/education/DMP/voting_b.html.

179 *would have turned the election*: "United States Presidental Election, 2000," Wikipedia, accessed August 13, 2015, https://en.wikipedia.org /wiki/United_States_presidential_election,_2000.

181 *"intended for honest men"*: Quoted in Aki Lehtinen, "The Borda Rule is Also Intended for Dishonest Men," *Public Choice* 133 (2007): 73–90, doi:10.1007/s11127-007-9178-5.

181 *the* Gibbard-Satterthwaite theorem: Allan Gibbard, "Manipulation of Voting Schemes: A General Result," *Econometrica* 41, no. 4 (1973):

587–601, doi:10.2307/1914083; and Mark Allen Satterthwaite, "Strategy-Proofness and Arrow's Conditions: Existence and Correspondence Theorems for Voting Procedures and Social Welfare Functions," *Journal of Economic Theory* 10, no. 2 (1975): 187–217, doi:10.1016/0022-0531(75)90050-2.

182 *Gibbard's two proposed solutions*: Allan Gibbard, "Manipulation of Schemes that Mix Voting with Chance," *Econometrica* 45, no. 3 (1977): 665–81.

Epilogue: Leaving the Nest

187 *"the pleasure of seeing it"*: Adam Smith, *The Theory of Moral Sentiments* (London and Edinburgh: A. Millar, A. Kincaid, and J. Bell, 1759), quoted in Joan B. Silk and Bailey R. House, "Evolutionary Foundations of Human Prosocial Sentiments," *PNAS* 108, suppl. 2 (2011): 10910–17, doi:10.1073/pnas.1100305108/-/DCSupplemental.

187 *gave way to progress*: Amanda Foreman, "Why Footbinding Persisted in China for a Millennium," *Smithsonian* magazine, February 2015, www.smithsonianmag.com/history/why-footbinding-persisted-china-millennium-180953971/?no-ist=&page=1.

188 *"hold human groups together"*: Cristina Bicchieri, *The Grammar of Society: The Nature and Dynamics of Social Norms* (New York: Cambridge University Press, 2006), ix.

188 *everyone else is wearing*: Bicchieri, 214.

188 *necessary for a society's existence*: Ibid.

189 *"maintenance of any public good"*: Bicchieri, 234.

189 *in a self-serving way*: Ilaria Castelli, Davide Massaro, Cristina Bicchieri, Alex Chavez, and Antonella Marchetti, "Fairness Norms and Theory of Mind in an Ultimatum Game: Judgments, Offers, and Decisions in School-Aged Children," PLoS ONE 9, no. 8 (2104): e105024, http://journals.plos.org/plosone/article?id=10.1371/journal.pone.0105024.

190 *We extend cooperation to strangers*: Bailey R. House, Joan B. Silk, Joseph Henrich, H. Clark Barrett, Brooke A. Scelza, et. al., "Ontogeny of Prosocial Behavior across Diverse Societies, *PNAS* 110 (2013): 14586–91.

Acknowledgments

Both of us talked and talked with scholars, friends, and family about this book. Here we would like to take the time to thank as many of them as we can remember.

First and last, our editor, Amanda Moon, introduced us to each other and helped to shape the development of this book in many ways, occasionally using a few game-theory tricks to manage her unruly authors. She and her assistant, Laird Gallagher, contributed important and helpful suggestions. Annie Gottlieb, our favorite copy editor, drove us to exhaustion, for which we are grateful, although we might not have said so at the time. We are also grateful to Debra Helfand, Lenni Wolff, Lottchen Shivers, and Scott Borchert at FSG, and to Mariette DiChristina, editor in chief and senior vice president of *Scientific American*. The idea for this book began with Jennifer Breheny Wallace when she interviewed Kevin for an article in *The Wall Street Journal* on game theory and parenting. And we would like to thank her for the inspiration.

Kevin pestered many of his game theorist friends to provide him with ideas and pointers. He would like to thank Carl Bergstrom, Cristina Bicchieri, Simon Huttegger, John Miller, Cailin O'Connor, Scott Page, Teddy Seidenfeld, Brian Skyrms, Rory Smead, Julia Staffel, and Elliott Wagner for many helpful discussions. Kevin also shamelessly milked his friends who offered up many more funny stories than we could fit in this book. Kevin would like to thank them all for helping out. In some cases, names were changed to protect the less-than-innocent.

Kevin's sister, Kim Rendfeld, an author herself, provided helpful writing guidance, as well as one or two reminders about Kevin's childhood. Before this project, Kevin's style of writing was as clear and captivating as you would expect from a philosopher—it wasn't. Paul's helpful advice and guidance nurtured in Kevin a sense of humor and storytelling that he didn't know he had. Kevin would like to thank Paul for his help, good humor, and patience.

Kevin's longtime partner, Korryn Mozisek, was a central contributor to the book in many ways. In addition to providing stories of her own, she listened to thousands of bad ideas and read several drafts. She is supportive and tolerant, and she shoulders burdens that only the spouse of an academic can truly understand. Kevin is lucky and very grateful.

Two very early contributors to this book were Jackie Spears and Dean Zollman—Kevin's parents. Kevin remembered his childhood as nothing but good behavior and kind thoughts. His parents were happy to take time out of their busy schedules to correct his error. As you have probably figured out, Kevin caused his parents much grief. They handled it with aplomb and sometimes, without knowing it, a little game theory. From an early age they encouraged Kevin to always ask questions about the world around him that instilled in him a deep curiosity. We all owe our existence to our parents, but Kevin owes his very much more. Books alone cannot express his gratitude.

Paul would like to thank the many writers and editors who over the years have helped him—and are still helping him—learn something about how to use the English language and how to report and tell a story. Although he didn't recognize it until years later, one of his best English teachers was Father James E. Farrell, a Jesuit priest who, in high school, taught Paul the importance of precision and clarity in language. Father Farrell asked questions about the Grecian urn and Ozymandias that Paul is still puzzling over.

His colleagues in the Invisible Institute, a writers' group in New York City, provided tremendous support and encouragement. Writing a book is a difficult business. If writers didn't get advances from their publishers that bound them to deliver, many books would probably never get written. The

Acknowledgments

Invisibles know what that feels like, and so it's good to be able to talk to them. And many thanks to the staff and to the bricks-and-mortar of the Writers Room, where Paul gets more done than anywhere else.

Paul's five children are among his most willing and persistent teachers, presumably because they believe he still has so much to learn. Paul's parents were also among his best teachers. He remembers his mother coming home from the public library with tall stacks of books, which is probably why he is producing his own stack of books now.

Paul would also like to thank Kevin, who proved to be a smart and capable tutor, and a careful, clear writer. Combine that with his good sense of humor, and it was inevitable that Paul and Kevin would become friends as well as collaborators. And many thanks, as always, to Paul's agent, Beth Vesel—our biggest fan.

Most of all, Paul would like to thank his wife, Elizabeth, who is not only the smartest editor he knows (apologies, Amanda) but his friend, his support, and his collaborator in everything. That includes parenting, where our reading and writing about game theory have helped us dare to believe that we might just get something right now and then.

Index

Index

Index

Index

Index

A NOTE ABOUT THE AUTHORS

PAUL RAEBURN is the award-winning author of four books, including *Do Fathers Matter?*, a National Parenting Publications Gold Medal winner. His articles have appeared in *Discover*, *The Huffington Post*, *The New York Times Magazine*, *Scientific American*, and *Psychology Today*, among many other publications. Follow him on Twitter (@praeburn) and visit his website at www.paulraeburn.com.

KEVIN ZOLLMAN is a game theorist and an associate professor of philosophy at Carnegie Mellon University. His work has been covered in *The Wall Street Journal*, *The New Yorker*, *Scientific American*, and elsewhere. Follow him on Twitter (@Kevin Zollman) and visit his website at www.kevinzollman.com.